Lecture Notes in Control and Information Sciences 196

Editor: M. Thoma

Stefano Battilotti

Noninteracting Control with Stability for Nonlinear Systems

Springer-Verlag London Ltd.

Series Advisory Board

Author

Stefano Battilotti, PhD
Dipartimento di Informatica e Sistemistica, Università degli Studi di Roma,
"La Sapienza", Via Eudossiana 18, 00184 Rome, Italy

ISBN 978-3-540-19891-8 ISBN 978-3-540-39336-8 (eBook)
DOI 10.1007978-3-540-39336-8

British Library Cataloguing in Publication Data
A catalogue record for this book is available from the British Library

Typesetting: Camera ready by author

69/3830-543210 Printed on acid-free paper

to Elisabetta and to my parents

CONTENTS

ACKNOWLEDGEMENTS

First of all, I wish to express my endless gratitude to Prof. Alberto Isidori for introducing me in the field of nonlinear control systems and for pointing me out, day after day, what a *good* researcher is. His few pieces of advice are worth many of the ideas, which every day come up to my mind.

In addition, I want to express my gratitude to all the contributors to the problem of noninteracting control, without whom I would have maybe ignored even what noninteraction is. I wish also to thank Prof. W.P.Dayawansa for patiently teaching me new methodologies and Prof. Jessy Grizzle for clarifying discussions and many stimulating ideas.

Finally, I wish to thank Proff. Alberto Isidori, Salvatore Monaco and Claudio Gori Giorgi for giving me the possibility of obtaining financial support by MURST, ASI and IBM, respectively.

CHAPTER 0

INTRODUCTION

The problem of noninteracting control has been studied since the late sixties by several authors. The papers of Falb & Wolovich ([28]), Gilbert ([36]), Morse & Wonham ([58], [89]–[90]), Basile & Marro ([1]) and Wang ([88]) are cornerstones in the noninteracting control theory for linear systems. In [58], [89]–[90] and, independently, in [1]the problem of noninteracting control has been formulated and solved in the framework of linear geometry, using mathematical tools such as linear vector spaces and matrix theory. These tools have been successfully used to address the issue of internal stability: an exhaustive theory is contained in [91]. A more algebraic approach has been pursued in [31], [36] and [88], while in [44], [46] and [73] a fundamental role is played by mathematical tools such as the Laplace transform and transfer function analysis. In the algebraic framework, we cannot help mentioning the celebrated Silverman's structure algorithm ([74]) and its fundamental role in understanding the structure at infinity ([73]), which is intrinsically related to the problem of noninteracting control. Later, during the eighties, a combination of algebraic and geometric tools has been used to solve some of the problems left open in the late sixties ([24], [25], [28], [45] and [48]).

The first efforts to extend to nonlinear systems the noninteracting control theory, available for linear systems, peeped in only at the beginning of the seventies with the paper of Porter ([69]), followed by few others ([35], [75], [74]). In 1981, Isidori et al. ([52]) formulated and studied for the first time the problem of nonlinear noninteracting control in the framework of differential geometry and distributions, opening the way to a solution to this problem in the spirit of [1] and [91]. At the same time, Singh ([76] and [77]) extended the Silverman's structure algorithm to a nonlinear setting. In 1985, Fliess ([32]) formulated and studied the noninteracting control problem in the framework of differential algebra. One year later, Descusse and Moog ([26] and [27]) extended to nonlinear systems the results of [88] and, at the same time, Nijmeijer and Schumacher ([62]–[64]) extended the results contained in [82], by using differential geometry and distribution theory.

In all these papers, throughout a period of nearly twenty years, the problem of internal stability has never been considered for nonlinear systems. Only in 1986, Ha and Gilbert

([41] and [42]) opened the way to a number of interesting contributions in this direction. In 1988, Isidori and Grizzle ([54]) approached this problem by using simple mathematical tools in the framework of differential geometry, followed so far by Wagner ([78]–[79]), Battilotti ([2]–[6]) and Zhan et al. ([92]). Up to now, the nonlinear noninteracting control problem with stability is well–understood but further work must be done either in the case of systems with outputs partitioned into given blocks (*block–partitioned* outputs) or in the case that the state of the system is not available for feedback (*measurement feedback*). However, all the contributions to the nonlinear noninteracting control problem with stability are hidden and scattered in the literature and have never been collected in an organic and self–contained theory, as it was done in the book [91] for linear systems. These notes are an useful supplement to the standard texts on the nonlinear control theory and present a self–contained and exhaustive discussion of the state–of–the–art of the noninteracting control problem with stability. This discussion contains, as a particular case, the one developed in [91] for linear systems.

0.1 What is noninteraction?

Let us consider nonlinear systems of the form

$$\begin{aligned} \dot{x} &= f(x) + \sum_{j=1}^{m} g_j(x)u_j, \\ y_i &= h_i(x), \qquad i = 1,\ldots,\nu, \end{aligned} \tag{1}$$

with $x \in \mathbb{R}^n$, $u_j \in \mathbb{R}$, $j = 1,\ldots,m$, $f(x)$, $g_j(x)$, $j = 1,\ldots,m$, smooth n–vector functions and $h_i(x)$, $i = 1,\ldots,\nu$, smooth p_i–vector functions, defined on some open set of $\mathbb{R}^n$, containing the origin. Moreover, let $\sum_{i=1}^{\nu} p_i = p$, $u = (u_1 \ \cdots \ u_m)^T$, $G(x) = (g_1(x) \ \cdots \ g_m(x))$ and $h(x) = (h_1^T(x) \ \cdots \ h_m^T(x))^T$. The system (1) has m inputs and p outputs, which are partitioned into ν given blocks. Moreover, we will assume that $x_0 = 0$ is an equilibrium point for $f(x)$, i.e. $f(0) = 0$, and that $g_1(x),\ldots,g_m(x)$ are linearly independent vectors at each x in their domain.

We will say that the system (1) is *noninteractive* if there exists a block partition $u_{I_1},\ldots,u_{I_{\nu+1}}$ of the input vector u, with $I_i \subset \{1,\ldots,m\}$ and $I_i \cap I_j = \{\phi\}$ for $j \neq i$ and $i,j = 1,\ldots,\nu+1$, such that each component of y_i is influenced (or "controlled") only by the components of u_{I_i}.

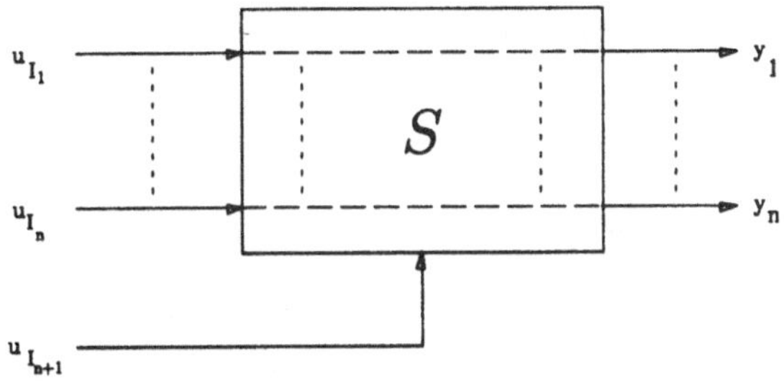

Figure 1.1

In other words, a system (1) is noninteractive if there exists a *disjoint* block partition of the input vector such that each component of the i–th output block is influenced only by the components of the i–th input block. Note that the $(\nu+1)$–th input block does not influence any output block (see Figure 1.1).

The property for which each component of y_i is influenced only by the components of u_{I_i} can be rigourously expressed in the following way. For each initial state $\bar{x}$ and for each pair of input vectors $u^1(t)$ and $u^2(t)$, defined for $t \geq 0$ and such that the components of their i–th block (corresponding to the partition $u_{I_1}, \ldots, u_{I_{\nu+1}}$) are equal, namely $u_s^1(t) = u_s^2(t)$ for all $s \in I_i$, we have $y_i^1(t) = y_i^2(t)$, where $y_i^1(t)$ and $y_i^2(t)$ are the outputs of the i–th block of $y(t)$ corresponding to $\bar{x}$ and to the input vectors $u^1(t)$ and $u^2(t)$, respectively.

Moreover, we will say that (1) is *noninteractive with (local) stability* in x_0 if, besides being noninteractive, it is also (locally) asymptotically stable in x_0 (in the sense of Lyapunov). In what follows, by stability we will implicitly mean asymptotic stability at the origin of the state space.

If (1) is not noninteractive with stability, we can try to modify the behaviour of the system (1) through suitable control laws in such a way that the above properties are achieved for the closed–loop system. In a natural way, one requires that these control laws preserve both the affine structure of the system (the closed–loop system depends linearly on the input vector u) and its equilibrium point (the origin of the state space is still an equilibrium point of the closed–loop system). Assuming that $x(t)$ is available for feedback (*full information*), we will consider either *static state–feedback laws* of the form

$$\begin{aligned} &u = \alpha(x) + \beta(x)v, \qquad v \in I\!R^m, \\ &\alpha(0) = 0, \end{aligned} \tag{2}$$

or *dynamic state-feedback laws* of the form

$$\begin{aligned} u &= \alpha(x,w) + \beta(x,w), && v \in \mathbb{R}^m, \\ \dot{w} &= \gamma(x,w) + \delta(x,w)v, && \\ \alpha(0,0) &= 0,\ \gamma(0,0) = 0, && w \in \mathbb{R}^{n^w}, \end{aligned} \tag{3}$$

with $v \in \mathbb{R}^m$, $w \in \mathbb{R}^{n^w}$ and $n^w \geq 0$. If $n^w = 0$, we obtain *static* feedback laws. In the case that $x(t)$ is not available (*measurement feedback*), we will consider *output-feedback* laws

$$\begin{aligned} u &= \alpha(h(x),w) + \beta(h(x),w)v, \\ \dot{w} &= \gamma(h(x),w) + \delta(h(x),w)v, \\ \alpha(0,0) &= 0,\ \gamma(0,0) = 0, \end{aligned} \tag{4}$$

The control laws of the form (2) are said to be *regular* if the matrix $\beta(x)$ is invertible at each x of an open and dense subset of $\mathbb{R}^n$. More generally, the control laws of the form (3) are said to be *invertible* if the system

$$\begin{aligned} \dot{x} &= f(x) + G(x)\alpha(x,w) + G(x)\beta(x,w)v, \\ \dot{w} &= \gamma(x,w) + \delta(x,w)v, \\ u &= \alpha(x,w) + \beta(x,w)v \end{aligned}$$

with input vector v and output vector u, is invertible in the sense of [77] (see section **2**.1.1 for further details). In particular, a regular feedback law is also invertible. Throughout these notes, we will consider only feedback laws which are invertible on an open and dense subset of $\mathbb{R}^{n+n^w}$ containing the origin.

0.2 Noninteraction with stability: a two–decade history

During the last two decades, necessary and sufficient conditions for which there exists a control law (2) (resp. (3)) such that the closed–loop system (1)–(2) (resp. (1)–(3)) is noninteractive (without stability) have been studied by several authors ([13], [15], [17]–[18], [20], [21], [24]–[32], [34]–[42], [44]–[55], [57]–[67], [69], [73], [75]–[78], [86]–[92]).

In the case that the closed–loop system is required to be also internally asymptotically stable, the problem has been completely solved for *linear* systems in [36], [58] and [89]–[90]. In [36], as far as *regular* feedback is concerned, it has been shown that every static feedback control law achieving noninteraction induces an internal dynamic, which has asymptotic properties completely independent from the control law itself. Thus, noninteraction and stability can be achieved via *static* feedback only if such dynamics is asymptotically stable (this dynamics will be denoted in the sequel by Σ_1). Assuming the plant controllable, a necessary and sufficient condition to achieve noninteraction with stability via *static* feedback is that Σ_1 be asymptotically stable and noninteraction itself can be achieved via *static* feedback.

More generally, one can use *dynamic* feedback in order to enlarge the class of linear systems for which noninteraction and stability can be achieved. In [58] and [89]–[90] it has been shown that, if the plant is controllable and *noninteraction* can be achieved via dynamic feedback, then also *noninteraction* and *stability* can be achieved via dynamic feedback. Thus, in the linear case and as far as *dynamic* feedback is concerned, the problem of achieving noninteraction and stability is *equivalent* to the problem of achieving *only* noninteraction.

In the case of *nonlinear* systems (1), the problem of achieving noninteraction and stability has been only recently addressed in [40]–[42] and [54]. In [40] and [54] it has been shown that, if it is possible to achieve noninteraction and stability via *regular* feedback, a suitable dynamics must be asymptotically stable. In the linear case, this dynamics is exactly Σ_1 and, thus, in the sequel it will be denoted in the same way. It has been also shown that, as far as *regular* feedback is concerned, if the system (1) has a controllable linear approximation, a necessary and sufficient condition for which there exists a *static* feedback law (2) such that (1)–(2) is noninteractive and its linear approximation is asymptotically stable is that the linear approximation of Σ_1 be asymptotically stable and that the system (1) can be rendered noninteractive via static feedback.

In the case $m = p = \nu$, i.e. each output controlled by one input, a characterization of all the possible regular feedback laws (2), which achieve noninteraction, has been given in [41] and [42]. However, the result is valid only for the class of *analytic* systems (1), i.e. systems in which $f(x)$ and $g_j(x)$ are analytic n–vector functions. Moreover, this characterization does not give any information about the class of regular feedback laws (2) for which (1)–(2) is also asymptotically stable.

At this point, in analogy with the linear case, it is natural to ask if the use of *dynamic* feedback can be of some help in the event that Σ_1 is unstable. A counterexample in [54] clearly shows that it is not possible, in general, to achieve noninteraction and stability via dynamic feedback, even if noninteraction itself can be achieved. The reason of this fact has been widely illustrated by Wagner in [86] and [87] for the class of analytic systems with $m = p = \nu$ for which noninteraction can be achieved via regular feedback or, equivalently, having *vector relative degree* defined at x_0. Wagner has shown that every *invertible* control law (3), achieving vector relative degree and noninteraction, induces an internal dynamics, which has asymptotic properties completely independent of the control law itself. Thus, also in this case, noninteraction and stability can be achieved only if this dynamics (which will be denoted in the sequel by Σ_2) is asymptotically stable. Moreover, as we expect, the dynamics Σ_2 is a *subdynamics* of Σ_1: this means that dynamic feedback can actually enlarge the class of systems (1) which can be rendered noninteractive and asymptotically stable.

The results shown in [86] and [87] have been extended in [92] to the class of systems (1) for which it is possible to obtain vector relative degree via *dynamic* feedback (the property of vector relative degree is preserved after applying *regular* feedback laws). It has been proved that every dynamic control law (3), which achieves vector relative degree, is such that (1)–(3) contains a dynamic *canonical* extension of (1), in the sense that (1)–(3) is *equal* to this extension up to local changes of coordinates and invertible dynamic control laws which achieve vector relative degree. Moreover, this canonical extension has itself vector relative degree and, thus, the dynamics Σ_2, correspondingly defined for it, must be asymptotically stable, if noninteraction and stability can be achieved for (1).

In the linear case, as expected, the dynamics Σ_2 is trivial or, which is the same, the problem of achieving noninteraction and stability via dynamic feedback is equivalent to the problem of achieving only noninteraction via dynamic feedback.

In [2]–[6] most of the problems, left open in the above papers, have been solved at least for the case $p = \nu$, i.e. each output controlled by one or more inputs. In particular, a complete set of necessary and sufficient conditions is now available to solve the noninteracting control problem with stability through *invertible* feedback laws, defined in a neighbourhood of the origin of the state space (*local* problem). As it will be clear, in the case $m = p = \nu$ the class of *invertible* feedback laws arises quite naturally when considering either *static* or *dynamic* feedback; for general systems, this constraining the class of feedback laws is *not* natural but allows to obtain significant and pictorial results. Some interesting results are also available either in the case of block–partitioned outputs (namely, either $p > \nu$) or in the case that the control law is defined everywhere in the state space (*global* problem) or when only the output of the system is available for feedback.

0.3 Scanning the contents

In chapter **1**, some background material is shortly reviewed for reader's convenience. The interested reader is referred for more details to the standard texts [7], [43] and [55].

In chapter **2** the problem of achieving noninteraction and stability for *linear* systems is revisited in such a way to highlight the main features in common with the nonlinear case (discussed in the following chapters). This problem is first formulated in terms of linear geometry: practically, one requires the existence of a feedback law (2) (resp. (3)) such that the closed–loop system (1)–(2) (resp. (1)–(3)) is noninteractive, asymptotically stable and the i–th output block is *controllable* through the i–th input block. The controllability of each i–th output block ensures that, by manipulating the inputs of the i–block, it is possible to reach an arbitrary value of the outputs of the i–th block. A main result of this section is that, under the hypothesis that the plant is controllable, the problem of achieving noninteraction with stability is equivalent to the problem of achieving *only* noninteraction. The proof, given in section **2**.3, proposes a constructive procedure slightly different from the one given in [91] and, as such, interesting.

In chapter **3**, we introduce the study of noninteracting control for *nonlinear* systems. In section **3**.1 some remarks on the class of feedback laws considered are discussed. In section **3**.2 we formulate the problem of achieving noninteraction and *local* stability in the case $m = p = \nu$ (*local* problem): practically, one looks for a feedback law (2) (resp. (3)), defined in a neighbourhood of x_0 (resp., of $(x_0^T, w_0^T)^T$), such that the closed–loop system (1)–(2) (resp. (1)–(3)) is noninteractive, asymptotically stable in x_0 (resp., in $(x_0^T, w_0^T)^T$) and has vector relative degree defined in x_0 (resp., in $(x_0^T, w_0^T)^T$). The last requisite ensures that the i–th output is effectively influenced or "controlled" by the i–th input. According to the structure of the control law, we have two kinds of problems: noninteracting control problem with local stability via *static* feedback (NLSS) and noninteracting control problem with local stability via *dynamic* feedback (NLSD).

Necessary and sufficient conditions to solve the problem of noninteracting control with local stability via *regular* feedback are discussed in section **3**.3. These conditions require (besides the possibility of achieving noninteraction and stability via regular feedback) the asymptotic stability of Σ_1 and the stabilizability via static feedback of suitable systems Σ_{i1}, $i = 1, \ldots, m$. Moreover, they extend the conditions given in [54], which, assuming the controllability of the linear approximation of (1), implicitly require the linear approximation of the systems Σ_{i1}, $i = 1, \ldots, m$, to be controllable.

Necessary and sufficient conditions are discussed also for the problem of achieving noninteraction with local stability via *invertible* feedback in the case $m = p = \nu$ (section **3**.4). As an intermediate result, the problem is first solved for the class of systems which can be rendered noninteractive via regular feedback (section **3**.4.2). A main result is that the noninteracting control problem with local stability via invertible dynamic feedback is solvable if and only if the dynamics Σ_2 is asymptotically stable and certain systems Σ_{i2}, $i = 1, \ldots, m$, are stabilizable via dynamic feedback. The proof of this important fact gives a step–by–step procedure to construct a solution to the problem. In a natural way, the conditions for which the problem of achieving noninteraction with local stability via *regular* feedback has solution imply the conditions for achieving noninteraction and stability via *invertible* feedback. As a final step, by using the results contained in [92], the above conditions are extended in a natural way to the case of systems (1) for which vector relative degree can be achieved via dynamic feedback (section **3**.4.4).

The possibility of reducing the dimension n^w of the dynamic compensator is considered in section **3**.4.3. By strenghening the hypotheses, it is possible to reduce this dimension down to a value, which in some cases is *minimal*.

In section **3**.4.5 we give some illustrative examples. In particular, the example of a noninteractive system with $m = p = 2$, which is not asymptotically stable. For this system Σ_1 is unstable and, thus, it is not possible to achieve noninteraction and stability via regular feedback. Nonetheless, this system satisfies the necessary and sufficient conditions to achieve noninteraction with local stability via invertible dynamic feedback. Moreover, in connection with this example, some remarks about the existence of *continuous* dynamic control laws, which solve the local problem, and the reduction of the dimension n^w of the dynamic compensator are discussed.

In chapter **4** the *global* problem is considered in the case $m = p = \nu$: practically, one looks for a feedback law (2) (resp., (3)), defined in $I\!R^n$ (resp., in $I\!R^{n+n^w}$), such that the closed–loop system (1)–(2) (resp. (1)–(3)) is noninteractive, globally asymptotically stable in x_0 (resp. in $(x_0^T, w_0^T)^T$) and has *uniform* vector relative degree, namely, vector relative degree $\{r_1, \ldots, r_m\}$ at each $x \in I\!R^n$. The analysis is restricted to the class of systems which have uniform relative degree and admit a suitable global decomposition (section **4**.3 and **4**.4). Some necessary and/or sufficient conditions (with illustrative examples) are discussed. These ones consist of a set of conditions which are exactly the same as those needed to solve the *local* problem plus some technical assumptions which ensure the trajectories of the closed–loop system (1)–(2) (resp. (1)–(3)) to be *bounded*. Moreover, in section **4**.5 the problem of achieving noninteraction with stability on arbitrary compact sets is considered.

In chapter **5** we consider the problem of noninteracting control with *local* stability in the case of *more* inputs than outputs.

First, we discuss the problem of achieving noninteraction with local stability via regular feedback and give some necessary and/or sufficient conditions in terms of suitable *generalized* systems Σ_1 and Σ_{i1}, $i = 1, \ldots, \nu$ (section **5**.1). One has that (besides the well–known conditions for achieving noninteraction via regular feedback) the asymptotic stability of Σ_1 plus the stabilizability via static feedback of Σ_{i1}, $i = 1, \ldots, \nu$, and that of a suitable system Σ_0 are sufficient to achieve noninteraction and stability via regular feedback. Under the assumption that Σ_0 is a trivial system, the previous conditions become also necessary. It is worth noting that, when $m = p = \nu$, the system Σ_0 is indeed trivial.

Similarly, in section **5**.2.1 and **5**.2.2 we discuss some necessary and/or sufficient conditions in terms of suitable *generalized* systems Σ_2 and Σ_{i2}, $i = 1, \ldots, \nu$, for achieving noninteraction with local stability via *invertible* feedback. We consider the class of systems for which noninteraction can be achieved via *regular* feedback. A main result is that if there exists an invertible feedback law (3) which achieves noninteraction and stability for the closed–loop system (1)–(3), then the dynamics Σ_2 must be asymptotically stable and the systems Σ_{i2}, $i = 1, \ldots, \nu$, must be stabilizable via dynamic feedback. On the other hand, these conditions, together with the hypothesis that Σ_0 is stabilizable via dynamic feedback, become also sufficient.

In the linear case, the stabilizability of the systems Σ_{i1}, Σ_{i2}, $i = 1, \ldots, \nu$, and Σ_0 is always ensured by the controllability of (1) and by their very definition.

Finally, in section **5**.2.3 we consider the case of *block–partitioned* outputs. In the spirit of section **2**.1 and unlike section **3**.2, the noninteracting control problem with local stability is formulated in terms of *controllability distributions*. Indeed, it is not possible to use the same formulation as given in section **3**.2, since in the case of block–partitioned outputs the concept of vector relative degree is meaningless. Practically, one requires the existence of a feedback law (2) (resp. (3)), defined in a neighbourhood of x_0 (resp., of $(x_0^T, w_0^T)^T$), such that the closed–loop system (1)–(2) (resp. (1)–(3)) is noninteractive, asymptotically stable in x_0 (resp., in $(x_0^T, w_0^T)^T$) and the i–th output block is locally *controllable* (from x_0 or, respectively, from $(x_0^T, w_0^T)^T$) through the inputs of the i–th block, according to a well–known definition ([56]–[60]).

In chapter **6**, we consider the problem of achieving noninteraction and stability via *output*–feedback. Some examples and counterexamples are given to illustrate the possible obstructions to the solution of this problem. Moreover, some necessary and some sufficient conditions for achieving noninteraction via *regular* feedback (section **6**.1) are given. However, the problem of finding necessary and sufficient conditions for achieving noninteraction and stability via output–feedback is still open and is a challenging object for further research.

CHAPTER 1

SOME BACKGROUND MATERIAL FROM DIFFERENTIAL GEOMETRY AND STABILITY

In this chapter, we will shortly review some basic mathematical tools and algorithms, which will be freely used throughtout these notes. The reader is supposed to be already familiar with the concepts of smooth vector field, covector field, distribution, codistribution, etc. etc. (see [7] or [55] for an easy introduction).

For a given distribution $\Delta \subset \mathbb{R}^n$, we can define the following codistribution

$$\Delta^{\perp}(x) = \{w \in (\mathbb{R}^n)^* : \langle w, d \rangle = 0, d \in \Delta(x)\},$$

where $(\mathbb{R}^n)^*$ is the dual of $\mathbb{R}^n$ and $\langle \cdot, \cdot \rangle$ is the standard inner product of $\mathbb{R}^n$. On the other hand, for a given codistribution $\Omega \subset (\mathbb{R}^n)^*$, we will define the following distribution

$$\ker\{\Omega\}(x) = \{d \in \mathbb{R}^n : \langle w, d \rangle = 0, w \in \Omega(x)\}.$$

In general, $\Delta^{\perp}$ and $\ker\{\Omega\}$ are not smooth. If Ω and Δ have constant dimension, then also $\Delta^{\perp}$ and $\ker\{\Omega\}$ are smooth and

$$\ker\{\Delta^{\perp}\} = \Delta\ , \ (\ker\{\Omega\})^{\perp} = \Omega.$$

Let U be a neighbourhood of x_0.

A smooth distribution $\Delta \in \mathbb{R}^n$ is said to be (locally) *involutive* if $[d_1, d_2] \in \Delta$ for all $d_1, d_2 \in \Delta$ and for all $x \in \mathbb{R}^n$ (respectively, for all $x \in U$), with $[\cdot, \cdot]$ denoting the Lie bracket of two vector fields.

A smooth distribution Δ is said to be (locally) *invariant* with respect to a given set of smooth vector fields $d_1, \dots, d_s \in \mathbb{R}^n$ if $[d_j, \tau] \in \Delta$ for $j = 1, \dots, s$, for all $x \in \mathbb{R}^n$ (respectively, for all $x \in U$) and for all $\tau \in \Delta$.

The smallest (locally) invariant distribution with respect to a given set of vector fields $d_1, \dots, d_s$ and containing a given smooth distribution Δ always exists, since the family of distributions satisfying the above properties is closed under the sum of distributions. This distribution will denoted by $\langle d_1, \ldots, d_s | \Delta \rangle$. Moreover, it can be computed, under certain regularity assumptions, in the following way. Let $[d_j, \Delta] = \{[d_j, \tau] : \tau \in \Delta\}$, $j = 1, \ldots, s$, and define the following sequence of distributions

$$\Delta_0 = \Delta$$

$$\Delta_k = \sum_{j=1}^{s} [d_j, \Delta_{k-1}] + \Delta_{k-1}.$$

On an open and dense subset $U^* \subset \mathbb{R}^n$, i.e. the set of points at which the distributions Δ_k, $n \geq k \geq 0$, have constant dimension, there exists $k^* < n$ such that

$$\Delta_{k^*} = \Delta_{k^*-1} = \langle \tau_1, \ldots, \tau_s | \Delta \rangle.$$

A smooth distribution Δ is said to be (locally) *controlled invariant* for (0.1) if there exists a feedback law (0.2), defined for all $x \in \mathbb{R}^n$ (respectively, for all $x \in U$), such that for each $x \in \mathbb{R}^n$ (respectively, for each $x \in U$) we have

$$[\tilde{f}, \Delta] \subset \Delta,$$

$$[\tilde{g}_j, \Delta] \subset \Delta, \qquad j = 1, \ldots, m,$$

with $\tilde{f} = f + G\alpha$ and $\tilde{g}_j = G\beta_j$, $j = 1, \ldots, m$, where β_j is the j-th column of β. In what follows, we will consider only *regular* feedback laws (2).

It is possible to show that a given smooth distribution is *locally* controlled invariant for (0.1) if and only if

$$[f, \Delta] \subset \Delta + \mathcal{G}$$

$$[g_j, \Delta] \subset \Delta + \mathcal{G} \qquad j = 1, \ldots, m.$$

This condition is no more sufficient for the existence of a feedback law (2), defined for *all* $x \in \mathbb{R}^n$, such that Δ is invariant under $\tilde{f}, \tilde{g}_1, \ldots, \tilde{g}_m$ (see [20] and [21]).

Let $\varphi(x)$ be a smooth (possibly vector-valued) function, defined on U. Clearly, span $\{dh\}$ is a codistribution defined on U. Under suitable regularity assumptions, it is possible to compute the maximal locally controlled invariant distribution for (0.1) contained in $\ker\{\text{span}\{d\varphi\}\}$, which in what follows will be denoted by $\mathcal{V}^*(\ker\{\text{span}\{d\varphi\}\})$. Toward this end, denoting by $L_\tau \omega$ the Lie derivative of the covector field ω along the vector field τ, for a given smooth codistribution Ω let $L_f \Omega = \{L_f \omega : \omega \in \Omega\}$ and $L_{g_j} \Omega = \{L_{g_j} \omega : \omega \in \Omega\}$ for $j = 1, \ldots, m$ and define the following sequence of codistributions

$$\Omega_0 = \text{span}\{d\varphi\}$$

$$\Omega_k = L_f(\Omega_{k-1} \cap \mathcal{G}^\perp) + \sum_{j=1}^{m} L_{g_j}(\Omega_{k-1} \cap \mathcal{G}^\perp) + \Omega_{k-1}.$$

The codistributions Ω_k and $\Omega_k \cap \mathcal{G}^\perp$, $n \geq k \geq 0$, have constant dimension in a suitable open and dense subset $U^* \subset \mathbb{R}^n$ and there exists $k^* < n$ such that $\Omega_{k^*} = \Omega_{k^*-1} = \mathcal{V}^*(\ker\{\text{span}\{d\varphi\}\})$ for all $x \in U^*$.

A given smooth distribution Δ is said to be a (local) *controllability distribution* for (0.1) if it is (locally) involutive and there exists a feedback law (2), defined for all $x \in \mathbb{R}^n$ (respectively, for all $x \in U$), and a subset $I \subset \{1, \ldots, m\}$ such that for all $x \in \mathbb{R}^n$ (respectively, for all $x \in U$) one has

$$\Delta = \langle \tilde{f}, \tilde{g}_1, \ldots, \tilde{g}_m | \text{span}\{\tilde{g}_i : i \in I\} \rangle.$$

Now, assume that the codistributions Ω_k and $\Omega_k \cap \mathcal{G}^\perp$ have constant dimension for all $x \in \mathbb{R}^n$ (respectively, for all $x \in U$). Also in this case, under some additional regularity assumptions, it is possible to compute the maximal (local) controllability distribution for (0.1) contained in $\ker\{\text{span}\{d\varphi\}\}$, which will be denoted in what follows by $\mathcal{R}^*(\ker\{\mathit{span}\{d\varphi\}\})$. Toward this end, let us define the following sequence of distributions

$$\mathcal{S}_0 = \mathcal{V}^*(\ker\{\mathit{span}\{d\varphi\}\}) \cap \mathcal{G}$$

$$\mathcal{S}_k = \mathcal{V}^*(\ker\{\mathit{span}\{d\varphi\}\}) \cap ([f, \mathcal{S}_{k-1}] + \sum_{j=1}^{m} [g_j, \mathcal{S}_{k-1}] + \mathcal{G}) + \mathcal{S}_{k-1}.$$

The distributions $\mathcal{S}_k$, $n \geq k \geq 0$, have constant dimension for all x in a suitable open and dense subset $U^* \subset \mathbb{R}^n$ and there exists $k^* < n$ such that $\mathcal{S}_{k^*} = \mathcal{S}_{k^*-1} = \mathcal{R}^*(\mathit{ker}\{\text{span}\{d\varphi\}\})$ for all $x \in U^*$.

The reader is supposed to know also the concepts of smooth manifold and tangent space at a point of a smooth manifold (see [7] or [55]). We will denote by $\mathcal{L}$ a given smooth manifold and by $T_x\mathcal{L}$ its tangent space at x. If $\mathcal{L} = \mathbb{R}^n$, its tangent space at x can be identified with $\mathbb{R}^n$ itself. Moreover, the tangent space at x of a *submanifold* of $\mathbb{R}^n$ can be identified with a *subspace* of $\mathbb{R}^n$.

For a given smooth function $\Phi(x)$, defined for all $x \in \mathbb{R}^n$ (resp. for all $x \in U$) and with values in $\mathbb{R}^n$, we will say that two smooth vector fields $v_1, v_2 \in \mathbb{R}^n$ are (locally) *Φ-correlated* if $(\Phi_{*x})_x v_1(x) = v_2(\Phi(x))$, for all $x \in \mathbb{R}^n$ (resp. for all $x \in U$), where Φ_{*x} is the differential of Φ at x.

A smooth mapping $z = z(x)$, defined for $x \in \mathbb{R}^n$ (respectively, for all $x \in U$) and with values in $\mathbb{R}^n$, is said to be a global (respectively, local) *diffeomorphism* (or *change of coordinates*) if

a) z is globally (respectively, locally) invertible,

b) z and z^{-1} are smooth mappings.

Two given systems Σ_1 and Σ_2, $\dot{x}^i = f^i(x^i) + g^i(x^i)u$, $i = 1, 2$ with f^1, f^2, g^1 and g^2 smooth vector fields, defined on a smooth submanifold $\mathcal{L}^1 \subset \mathbb{R}^n$ and $\mathcal{L}^2 \subset \mathbb{R}^n$, respectively, are said to be (locally) *diffeomorphic* if there exists a (local) diffeomorphism $\Phi : \mathcal{L}^1 \to \mathcal{L}^2$ such that the vector fields f^1 and f^2 and, respectively, g^1 and g^2 are (locally) Φ–correlated.

A smooth submanifold $\mathcal{L} \subset \mathbb{R}^n$ is said to be a *integral manifold* of a smooth distribution Δ if for each $x \in \mathcal{L}$ we have $T_x\mathcal{L} = \Delta(x)$.

An integral manifold $\mathcal{L}$ of Δ is said to be *maximal* if it is connected and every other integral manifold of Δ is contained in $\mathcal{L}$. The set of maximal integral manifolds (or *leaves*) of Δ is said to be a *foliation* of Δ. The leaf of Δ passing through x will be denoted by $\mathcal{L}_x^\Delta$.

The previous definitions can be formulated *locally* if we replace $\mathbb{R}^n$ by an open neighbourhood of x_0.

Let $\Delta \subset \mathbb{R}^n$ be a smooth distribution, which is involutive, invariant with respect to $f(x), g_1(x), \ldots, g_m(x)$ and with constant dimension in U. By Frobenius' theorem (see [7], theorem 8.3) there exists a change of coordinates $z(x) = (\, z_1^T(x) \quad z_2^T(x)\,)^T$ (z_i is possibly a vector–valued function), defined in a neighbourhood U of x_0, such that (0.1) is expressed in these coordinates by

$$\dot{z}_1 = f_1(z) + \sum_{j=1}^{m} g_{1j}(z)u_j$$

$$\dot{z}_2 = f_2(z_2) + \sum_{j=1}^{m} g_{2j}(z_2)u_j,$$

with $\Delta = \text{span}\{\partial/\partial z_1\}$. The leaves of Δ are described by the level sets

$$\{z \in U : z_2 = cost\}.$$

Since $f(x_0) = 0$ and Δ is invariant with respect to f, it makes sense to consider the *restriction* of f to $\mathcal{L}_{x_0}^\Delta$. This restriction induces a well–defined dynamics evolving on $\mathcal{L}_{x_0}^\Delta$ and described by

$$\dot{z}_1 = f_1(z_1, 0).$$

Now, let us consider the system (0.1) with $\nu = p$, i.e. ν outputs. We will say that (0.1) has *(vector) relative degree* at x_0 ([55], pp. 234–235) if there exist integers $r_1, \ldots, r_p \geq 1$ such that

a) $L_{g_j} L_f^k h_i(x) = 0$ for $i = 1, \ldots, p$, $0 \leq k < r_i - 1$, $j = 1, \ldots, m$ and for all $x \in U$; $L_{g_j} L_f^{r_i - 1} h_i(x) \neq 0$ for some $j \in \{1, \ldots, m\}$ and for all $x \in U$;

b) the *decoupling matrix*, defined as

$$\begin{pmatrix} L_{g_1}L_f^{r_1-1}h_1(x) & \cdots & L_{g_m}L_f^{r_1-1}h_1(x) \\ \cdots & \cdots & \cdots \\ L_{g_1}L_f^{r_p-1}h_p(x) & \cdots & L_{g_m}L_f^{r_p-1}h_p(x) \end{pmatrix},$$

has rank p for all $x \in U$.

If the above definition holds with U replaced by $\mathbb{R}^n$, we will say that (0.1) has *uniform (vector) relative degree.*

Finally, we will shortly review some basic definitions from the stability theory (the reader is referred to [43] for a detailed and exhaustive discussion). Assume that the conditions for the existence of the solution of $\dot{x} = f(x,t)$ are satisfied. Moreover, assume that $f(0,t) = 0$ for all $t \geq 0$.

We will say that $\dot{x} = f(x,t)$ is *stable* in x_0 if, denoting by $x(t, t_0, \bar{x})$ the solution of $\dot{x} = f(x,t)$ at $t \geq t_0$, with initial condition $\bar{x}$ and initial time t_0, for each $\epsilon > 0$ there exists δ_{ϵ,t_0} such that for $\|\bar{x}\| < \delta_{\epsilon,t_0}$ we have $\|x(t,t_0,\bar{x})\| < \epsilon$ ($\|\cdot\|$ is the standard norm of $\mathbb{R}^n$).

If, in the above definition, δ depends *only* on ϵ, we will say that $\dot{x} = f(x,t)$ is *uniformly stable* (with respect to t_0) in x_0. If f does not depend explicitly on t, the stability is always uniform.

Moreover, we will say that x_0 is *locally attractive* for $\dot{x} = f(x,t)$ if there exists an open neighborhood U_{t_0} of x_0 such that, for all $\bar{x} \in U_{t_0}$, $x(t,t_0,\bar{x})$ is bounded and $\|x(t,t_0,\bar{x})\| \to x_0$ for $t \to \infty$. Moreover, if $U_{t_0} = \mathbb{R}^n$, then we will say that x_0 is *globally attractive* for $\dot{x} = f(x,t)$. The set of points for which x_0 is attractive is referred to as *basin of attraction* (of x_0).

If, in the above definition, U_{t_0} does not depend on t_0 and $\|x(t,t_0,\bar{x})\| \to x_0$ for $t \to \infty$ uniformly with respect to t_0, we will say that x_0 is *uniformly* attractive. If f does not depend on t, the attractivity of x_0 is always uniform.

Throughout these notes, we will be concerned with uniform stability and uniform attractivity and we will simply say stability and, respectively, attractivity.

We will say that $\dot{x} = f(x,t)$ is globally (respectively, locally) *asymptotically stable* in x_0 if x_0 is globally (respectively, locally) attractive and $\dot{x} = f(x,t)$ is stable in x_0.

The system (0.1) is said to be globally (respectively, locally) *asymptotically stabilizable* at x_0 via *static state-feedback* if there exists a control law $u = \alpha(x)$, $\alpha(x_0) = 0$, defined for all $x \in \mathbb{R}^n$ (respectively, for all $x \in U$), such that $\dot{x} = f(x) + G(x)\alpha(x)$ is globally (respectively, locally) asymptotically stable in x_0. Similarly, we can define the asymptotic stabilization either via *dynamic state-feedback* or via *dynamic output-feedback*.

By now, many interesting results are available to investigate the asymptotic stabilizability of a nonlinear system of the form (0.1). A first kind of test is given in terms of the matrices $(\frac{\partial f}{\partial x}(x_0), G(x_0))$, or, equivalently, in terms of the linear approximation of (0.1) in x_0. If this linear approximation is stabilizable (namely, the uncontrollable modes are asymptotically stable), then the nonlinear system (0.1) can be *locally* asymptotically stabilized through a *linear* state–feedback law ([18]). If some eigenvalues of $\frac{\partial f}{\partial x}(0)$ are not controllable and have zero real part (*critical case*), the theory of center manifold can be used in some simple cases to design stabilizing controllers ([8], [22], [23]). On the other hand, if some eigenvalues of $\frac{\partial f}{\partial x}(0)$ are not controllable and have *positive* real part, a smooth stabilizing feedback law cannot exist and we have to look for *continuous* feedback laws.

An important class of nonlinear systems, which can be always *locally* asymptotically stabilized through *static* state–feedback, is given by the systems which have *relative degree* and are *minimum phase* ([11], [12]). *Global* stabilization of this class of systems requires some additional assumptions (see [80], [81] and [82]).

Interesting results on global state–feedback stabilization have been also given in terms of *control Lyapunov functions* ([84], [85]). Necessary and sufficient conditions for state–feedback stabilization are available for two and three–dimensional analytic systems ([16], [22], [23]).

The following proposition gives an interesting result on *local* stability of triangular systems.

Proposition 1. ([55]) Let

$$\begin{aligned} \dot{x} &= f_1(x, z) \\ \dot{z} &= f_2(z), \end{aligned} \tag{1}$$

with $f_1(0,0) = 0$ and $f_2(0) = 0$. Moreover, assume that $\dot{z} = f_2(z)$ and $\dot{x} = f_1(x,0)$ are locally asymptotically stable in $x = 0$ and $z = 0$, respectively. Then, the overall system (1) is locally asymptotically stable in $(x, z) = (0,0)$. □

Let us denote by $x(t, 0, \bar{x}, u(t))$ the solution of (0.1) at time $t \geq 0$, starting at $t = 0$ and from $\bar{x}$ with input $u(t)$. The system (0.1) is said to be *convergent input bounded state* (CIBS) if, for each bounded $u(t)$ such that $||u(t)|| \to 0$ for $t \to \infty$ and for each $\bar{x} \in I\!R^n$, $x(t, 0, \bar{x}, u(t))$ is bounded and $||x(t, 0, \bar{x}, u(t))|| \to 0$ for $t \to \infty$.

The following proposition generalizes proposition 1 to a global setting.

Proposition 2. ([79]) Let us consider (1) and assume that $\dot{x} = f_1(x,0)$ and $\dot{z} = f_2(z)$ are globally asymptotically stable in $x = 0$ and $z = 0$, respectively, and that $\dot{x} = f_1(x, z)$ is convergent input bounded state. Then, the overall system (1) is globally asymptotically stable in $(x, z) = (0,0)$. □

CHAPTER 2

A REVIEW OF NONINTERACTING CONTROL WITH STABILITY FOR LINEAR SYSTEMS

2.1 Problem formulation

Let us consider the following linear system with m inputs and p block–partitioned outputs

$$\begin{aligned} \dot{x} &= Ax + \sum_{j=1}^{m} B_j u_j, \\ y_i &= C_i x, \qquad i = 1, \dots, \nu, \end{aligned} \tag{1.1}$$

where C_i is a $p_i \times n$ matrix and $x \in \mathbb{R}^n$. In what follows, we denote by span $\{d_1, \dots, d_s\}$ (or, simply, by span$\{D\}$, with $D = (\, d_1 \;\; \cdots \;\; d_s \,)$) the subspace of $\mathbb{R}^n$ generated by the vectors $d_1, \dots, d_s$ and by Ker W, where $W = (\, w_1^T \;\; \cdots \;\; w_s^T \,)^T$ and $w_1, \dots, w_s$ are linearly independent row vectors, the subspace of $\mathbb{R}^n$ generated by the vectors $v_{n-s}, \dots, v_n$ such that $\langle w_i, v_j \rangle = 0$ for $i = 1, \dots, s$ and $j = n-s, \dots, n$. Moreover, let $B = (\, B_1 \;\; \cdots \;\; B_m \,)$, $u = (\, u_1 \;\; \cdots \;\; u_m \,)^T$, $C = (\, C_1^T \;\; \cdots \;\; C_\nu^T \,)^T$, $\mathcal{K}_i = \text{Ker}\,\{C_i\}$ and $\mathcal{B} = \text{span}\{B\}$.

As already pointed out in the introduction, at the beginning of the seventies Gilbert and, independently, Morse and Wonham solved the problem of finding a static feedback law

$$u = Fx + Gv, \qquad v \in \mathbb{R}^m, \tag{1.2}$$

with G $m \times m$ invertible matrix (in this case, the feedback law is said to be *regular*), such that the system (1.1)–(1.2) is noninteractive and asymptotically stable. In a subsequent paper, Morse and Wonham gave a necessary and sufficient condition for the existence of a feedback law

$$\begin{aligned} u &= F_1 x + F_2 w + Gv, \qquad & v \in \mathbb{R}^m, \\ \dot{w} &= \Gamma_1 x + \Gamma_2 w + Hv, \qquad & w \in \mathbb{R}^{n^w}, \; n^w \geq 0, \end{aligned} \tag{1.3}$$

such that the system (1.1)–(1.3) is noninteractive and asymptotically stable. In this chapter, we will shortly review the solution to the problem of achieving noninteraction and stability for (1.1), keeping as close as possible to what will be the solution to the same problem for nonlinear systems.

Before going through the technical details, we spend few words on the basic assumptions. First, it is quite natural to assume that the pair (A, B) is controllable or, more generally, stabilizable. In the last case, one can think that the state variables of the uncontrollable dynamics of (1.1) go to zero sufficiently fast. Moreover, we assume that C_i is a full rank matrix, otherwise the problem of achieving noninteraction is unrealistic (see also [90] and [91]).

For reader's convenience, we will shortly recall some basic concepts of linear geometry, which allow a very elegant approach to the problem of achieving noninteraction and stability. Many of these concepts can be viewed in the more general context of differential geometry (see chapter **1**).

A subspace $\mathcal{R} \subset \mathbb{R}^n$ is said to be *invariant* under A if $A\mathcal{R} \subset \mathcal{R}$, where $A\mathcal{R} = \{v \in \mathbb{R}^n : v = Ar, r \in \mathcal{R}\}$. We denote by $\langle A|\text{span}\{d_1, \ldots, d_s\}\rangle$ (or, more simply, by $\langle A|\text{span}\{D\}\rangle$) the smallest subspace of $\mathbb{R}^n$ which is invariant under A and contains $\text{span}\{d_1, \ldots, d_s\}$. This subspace always exists, since the family of invariant subspaces is closed under intersection of subspaces, and can be computed in a finite number of steps by the algorithm given in the appendix. If a subspace $\mathcal{R} \subset \mathbb{R}^n$ is invariant under A, the restriction of A to $\mathcal{R}$, denoted by $A|\mathcal{R}$, induces a well-defined dynamics evolving on $\mathcal{R}$. If $\mathcal{R}$ is invariant under A, also the canonical projection $P(\mathbb{R}^n, \mathcal{R}) : \mathbb{R}^n \to \mathbb{R}^n/\mathcal{R}$, where $\mathbb{R}^n/\mathcal{R}$ is the quotient space $\mathbb{R}^n$ modulo $\mathcal{R}$, induces a well-defined dynamics through the linear map $\bar{A} : \mathbb{R}^n/\mathcal{R} \to \mathbb{R}^n/\mathcal{R}$ such that $P(\mathbb{R}^n, \mathcal{R})A = \bar{A}P(\mathbb{R}^n, \mathcal{R})$ (here, linear maps are identified with their matrix representation).

A subspace $\mathcal{R} \subset \mathbb{R}^n$ is said to be *controlled invariant* for (1.1) if there exists a feedback law $u = Fx$ such that $\mathcal{R}$ is invariant under $A + BF$.

A subspace $\mathcal{R} \subset \mathbb{R}^n$ is said to be a *controllability subspace* for (1.1) if there exists a feedback law $u = Fx + Gv$ and a column (block) partition G_1, G_2 of G such that $\mathcal{R} = \langle A + BF|\text{span}\{BG_1\}\rangle$. In particular, any controllability subspace is controlled invariant.

As well-known in the literature, a controllability subspace enjoys the following important property.

Proposition 1.1. ([91], theorem 5.1) *Let $\mathcal{R} \subset \mathbb{R}^n$ be a controllability subspace for (1.1). Then, there exist a feedback law $u = Fx + Gv$ and a column (block) partition G_1, G_2 of G such that*

$$\mathcal{R} = \langle A + BF|\text{span}\{BG_1\}\rangle$$

and $(A + BF)|\mathcal{R}$ is a Hurwitz matrix, i.e. its eigenvalues have negative real part. □

The basic argument behind the proof of proposition 1.1 is that, for any feedback law $u = Fx + Gv$ and for any column (block) partition G_1, G_2 of G such that $\mathcal{R} = \langle A + BF | \text{span}\{BG_1\}\rangle$, the pair $((A + BF)|\mathcal{R}, \Pi(\mathbb{R}^n, \mathcal{R})(BG_1))$, where $\Pi(\mathbb{R}^n, \mathcal{R}) : \mathbb{R}^n \to \mathcal{R}$ is the natural projection on $\mathcal{R}$ along any $\mathcal{S}$ such that $\mathcal{R} \oplus \mathcal{S} = \mathbb{R}^n$, is controllable. We remark that $\Pi(\mathbb{R}^n, \mathcal{R})(BG_1)$ does not depend on the choice of $\mathcal{S}$, since $BG_1 \in \mathcal{R}$.

At the beginning of the eighties, the controllability distributions were pointed out in [52] (and, independently, in [68]) as the natural substitutes of the controllability subspaces in many important control problems. However, although the controllability distributions inherit many properties of the controllability subspaces, proposition 1.1 cannot be extended, in general, to the controllability distributions, since for nonlinear systems controllability does *not* imply the existence of an even continuous feedback law, which asymptotically stabilizes the system at the equilibrium point (a very intuitive reason of this fact is given in [80]).

Now, we will show that the problem of achieving noninteraction is equivalent, in the framework of linear geometry, to the existence of ν controllability subspaces $\mathcal{R}_i \subset \bigcap_{j\neq i} \mathcal{K}_j$, $i = 1, \ldots, \nu$, which satisfy an *output controllability* property plus a *compatibility* property.

Toward this end, first assume that there exists a feedback law (1.2) such that the system (1.1)–(1.2) is noninteractive. Let $v_1, \ldots, v_{\nu+1}$ be the (block) partition of the input vector v for which (1.1)–(1.2) is noninteractive and $G_1, \ldots, G_{\nu+1}$ be the corresponding column (block) partition of G. Moreover, let $\widetilde{A} = A + BF$ and $\widetilde{B}_j = BG_j$, $j = 1, \ldots, \nu+1$. It can be easily shown that a necessary condition for (1.1)–(1.2) to be noninteractive is that

$$\langle \widetilde{A} | \text{span}\{\widetilde{B}_i, \widetilde{B}_{\nu+1}\}\rangle \subset \bigcap_{j\neq i} \mathcal{K}_j, \qquad i = 1, \ldots, \nu.$$

Note from the formula above that the feedback law (1.2) renders *simultaneously* invariant the controllability subspaces $\mathcal{R}_i = \langle \widetilde{A} | \text{span}\{\widetilde{B}_i, \widetilde{B}_{\nu+1}\}\rangle$, $i = 1, \ldots, \nu$ (this property is usually referred to as *compatibility* of the subspaces $\mathcal{R}_i$, $i = 1, \ldots, \nu$). In other words, there must exist ν controllability subspaces $\mathcal{R}_1, \ldots, \mathcal{R}_\nu$, a feedback law (1.2) and a column (block) partition $G_1, \ldots, G_{\nu+1}$ of G such that

$$\mathcal{R}_i \subset \bigcap_{j\neq i} \mathcal{K}_j, \qquad i = 1, \ldots, \nu, \tag{1.4}$$

$$\mathcal{R}_i = \langle \widetilde{A} | \text{span}\{\widetilde{B}_i, \widetilde{B}_{\nu+1}\}\rangle, \qquad i = 1, \ldots, \nu. \tag{1.5}$$

Conversely, if there exist ν controllability subspaces $\mathcal{R}_1, \ldots, \mathcal{R}_\nu$, a feedback law (1.2) and a column (block) partition $G_1, \ldots, G_{\nu+1}$ of G which satisfy (1.4) and (1.5), then (1.1)–(1.2) is noninteractive. In this case, we say that $\mathcal{R}_1, \ldots, \mathcal{R}_\nu$ is a *solution* of the problem of achieving noninteraction via static state–feedback (this solution is said to be *regular* if the feedback law (1.2) is regular).

Naturally, besides the noninteraction property, one requires also that the i–th output (block) of $\widetilde{\Sigma}$ be *controllable* through the inputs of the i-th block, namely, the possibility of achieving an arbitrary value of $y_i(t)$ in $\mathbb{R}^{p_i}$ by manipulating the inputs of the i–th block. This property can be immediately translated in geometric terms by $C_i\mathcal{R}_i = \mathbb{R}^{p_i}$ or, equivalently,

$$\mathcal{R}_i + \mathcal{K}_i = \mathbb{R}^n. \tag{1.6}$$

The above property is usually referred to as *output controllability* property.

If we consider the class of static state–feedback laws and motivated by the above characterization, we can formulate the problem of achieving noninteraction and stability as follows.

Noninteracting control with stability via static state–feedback (NSS). *Find, if possible, ν controllability subspaces $\mathcal{R}_1, \ldots, \mathcal{R}_\nu$ for (1.1), a feedback law (1.2) and a column (block) partition $G_1, \ldots, G_{\nu+1}$ of G which satisfy (1.4), (1.5) and (1.6) and such that (1.1)–(1.2) is asymptotically stable.*

If also dynamic state–feedback laws are considered, it can be easily shown that solving the problem of achieving noninteraction for (1.1) via dynamic state–feedback amounts to solving the same problem for a suitable *extended* system by using *nonregular* static state–feedback laws with a particular structure. Indeed, let us define the following *extended* system

$$\begin{aligned} \dot{x} &= Ax + Bu, \\ \dot{w} &= u^w, \qquad w \in \mathbb{R}^{n^w}, \; n^w \geq 0,, \\ y_i &= C_i x, \qquad i = 1, \ldots, \nu, \end{aligned}$$

where u^w is a vector of "fictitious" inputs. If

$$x^e = \begin{pmatrix} x \\ w \end{pmatrix}, \; u^e = \begin{pmatrix} u \\ u^w \end{pmatrix},$$

$$A^e = \begin{pmatrix} A & 0_{n \times n^w} \\ 0_{n^w \times n} & 0_{n^w \times n^w} \end{pmatrix}, \; B^e = \begin{pmatrix} B & 0_{n \times n^w} \\ 0_{n^w \times m} & I_{n^w \times n^w} \end{pmatrix},$$

$$C_i^e = \begin{pmatrix} C_i & 0_{p_i \times n^w} \end{pmatrix}, \; C^e = \left((C_1^e)^T \;\; \cdots (C_\nu^e)^T \right)^T,$$

where I is the identity matrix, 0 is the zero matrix and the subscripts denote the dimensions of these matrices, we can rewrite the *extended* system in the following form

$$\begin{aligned} \dot{x}^e &= A^e x^e + B^e u^e, \\ y_i &= C_i^e x^e, \qquad i = 1, \ldots, \nu. \end{aligned} \tag{1.7}$$

Moreover, let $\mathcal{B}^e = \text{span}\{B^e\}$ and $\mathcal{K}_i^e = \text{Ker}\,\{C_i^e\}$ (note that $\mathcal{B}^e$ and $\mathcal{K}_i^e$ are subspaces of $\mathbb{R}^{n+n^w}$). It follows immediately that noninteraction can be achieved for (1.1) via feedback laws (1.3) if (and only if) this property can be achieved for (1.7) via feedback laws with the following structure

$$u^e = F^e x^e + G^e v^e,$$

$$F^e = \begin{pmatrix} F_1 & F_2 \\ \Gamma_1 & \Gamma_2 \end{pmatrix}, \, G^e = \begin{pmatrix} G & 0_{n \times n^w} \\ H & 0_{n^w \times n^w} \end{pmatrix},$$

which are clearly *nonregular*. This also implies that achieving noninteraction for (1.1) is equivalent to finding n^w and ν compatible controllability subspaces $\mathcal{R}_i^e \subset \bigcap_{j \neq i} \mathcal{K}_j^e$, $i = 1, \ldots, \nu$ (in this case, $\mathcal{R}_1^e, \ldots, \mathcal{R}_\nu^e$ is a *nonregular* solution).

Motivated by the above facts, we can formulate the problem of achieving noninteraction and stability via dynamic state–feedback as follows. Let $G_1^e, \ldots, G_{\nu+1}^e$ be a column (block) partition of G^e and, correspondingly, let $\widetilde{A}^e = A^e + B^e F^e$ and $\widetilde{B}_j^e = B^e G_j^e$, $j = 1, \ldots, \nu + 1$.

Noninteracting control with stability via dynamic state–feedback (NSD). *Find, if possible, $n^w \geq 0$, ν controllability subspaces $\mathcal{R}_1^e, \ldots, \mathcal{R}_\nu^e$ for (1.7), a feedback law (1.3) and a column (block) partition $G_1^e, \ldots, G_{\nu+1}^e$ of G^e such that*

$$\mathcal{R}_i^e \subset \bigcap_{j \neq i} \mathcal{K}_j^e, \tag{1.8}$$

$$\mathcal{R}_i^e = \langle \widetilde{A}^e | \, span\{\widetilde{B}_i^e, \widetilde{B}_{\nu+1}^e\} \rangle, \tag{1.9}$$

$$\mathcal{R}_i^e + \mathcal{K}_i^e = \mathbb{R}^{n+n^w}, \tag{1.10}$$

and (1.1)-(1.3) is asymptotically stable.

In what follows, we will use the notations NS and ND to denote, respectively, the problem of noninteracting control (without internal stability) via static state–feedback and the problem of noninteracting control (without internal stability) via dynamic state–feedback. Moreover, if not otherwise stated, we consider only state–feedback laws.

2.2 Noninteracting control with stability via static state–feedback

To begin with, we will give necessary and sufficient conditions to solve NSS via regular feedback. This constraining the class of feedback laws considered has no practical reason but the fact that in this case a simple necessary and sufficient condition is available to solve NS, given by

$$\sum_{i=1}^{\nu} \mathcal{R}_i^* \cap \mathcal{B} = \mathcal{B} \tag{2.1}$$

(see [58] and [90]). Throughout this section, we will assume (2.1).

Let $\mathcal{R}_i^*$ be the maximal controllability subspace contained in $\bigcap_{j \neq i} \mathcal{K}_j$. This subspace always exists, since the family of controllability subspaces is closed under sum of subspaces, and can be computed in a finite number of steps by the algorithm given in chapter 1. Moreover, let $\mathcal{R}^* = \bigcap_{i=1}^{\nu} \sum_{j \neq i} \mathcal{R}_j^*$ and $\mathcal{Q}^*$ be the maximal controllability subspace contained in $\mathcal{R}^*$. Note that $\mathcal{Q}^* \subset \mathcal{R}_i^*$, $i = 1, \ldots, \nu$. As a matter of fact, $\mathcal{Q}^* \subset \mathcal{R}^*$ by definition. But $\sum_{j \neq i} \mathcal{R}_j^* \subset \mathcal{K}_i$ and, thus, $\mathcal{Q}^* \subset \mathcal{R}^* \subset \bigcap_{i=1}^{\nu} \mathcal{K}_i$. Since $\mathcal{R}_i^*$ is the maximal controllability subspace contained in $\bigcap_{j \neq i} \mathcal{K}_j$ and since $\mathcal{Q}^*$ is itself a controllability subspace contained in $\bigcap_{i=1}^{\nu} \mathcal{K}_i$, which, in turn, is contained in $\bigcap_{j \neq i} \mathcal{K}_j$, it follows that $\mathcal{Q}^* \subset \mathcal{R}_i^*$.

2.2.1 Invariant dynamics in noninteracting control via static feedback

Before giving the necessary and sufficient condition to solve NSS, we need introduce some linear dynamics which can be *uniquely* (in some sense) associated to a given system satisfying (2.1). As it will be clear later, these dynamics are crucial in solving NSS.

By standard arguments (see [58]), it can be shown that (2.1) is equivalent to $\mathcal{R}_1^*, \ldots, \mathcal{R}_\nu^*$ being a regular solution. As a consequence, there exist a regular feedback law (1.2) and a column (block) partition $G_1, \ldots, G_{\nu+1}$ of G such that $\mathcal{R}_i^* = \langle \widetilde{A} | \operatorname{span}\{\widetilde{B}_i, \widetilde{B}_{\nu+1}\} \rangle$, $i = 1, \ldots, \nu$.

For a given (block) partition $x_1, \ldots, x_s$ of the state vector x, with $x_i \in \mathbb{R}^{m_i}$, we will denote by $\frac{\partial}{\partial x_i}$ the $n \times m_i$ matrix having the i–th block (corresponding to the given partition of x) equal to $I_{m_i \times m_i}$ and the j–th block equal to $0_{m_j \times m_i}$ for $j \neq i$.

Since $\mathcal{R}_1^*, \ldots, \mathcal{R}_\nu^*$ are invariant under $\widetilde{A}$, also $\sum_{j \neq i} \mathcal{R}_j^*$, $i = 1, \ldots, \nu$, $\mathcal{R}^*$ and $\mathcal{Q}^*$ enjoy this property by their very definition. Thus, there exists a change of coordinates $z =$

$(z_1^T \quad \cdots \quad z_{\nu+2}^T)^T$ such that (1.1)–(1.2) is expressed in these coordinates by

$$
\begin{aligned}
\dot{z}_i &= \widetilde{A}_{ii} z_i + \widetilde{B}_{ii} v_i, \qquad i = 1, \dots, \nu, \\
\dot{z}_{\nu+1} &= \widetilde{A}_{\nu+1,1} z_1 + \dots + \widetilde{A}_{\nu+1,\nu+1} z_{\nu+1} + \sum_{i=1}^{\nu} \widetilde{B}_{\nu+1,j} v_j, \\
\dot{z}_{\nu+2} &= \widetilde{A}_{\nu+2,1} z_1 + \dots + \widetilde{A}_{\nu+2,\nu+2} z_{\nu+2} + \sum_{i=1}^{\nu+1} \widetilde{B}_{\nu+2,j} v_j, \\
y_i &= C_{ii} z_i, \qquad i = 1, \dots, \nu,
\end{aligned}
\tag{2.2}
$$

with $\sum_{j \neq i} \mathcal{R}_j^* = \operatorname{span}\{\partial/\partial z_j : j \neq i\}$, $i = 1, \dots, \nu$, $\mathcal{R}^* = \operatorname{span}\{\partial/\partial z_{\nu+1}, \partial/\partial z_{\nu+2}\}$ and $\mathcal{Q}^* = \operatorname{span}\{\partial/\partial z_{\nu+2}\}$. In what follows, for simplicity of notation, we let $z = x$ and $v = u$.

Now, let us consider the well–defined dynamics

$$
\dot{x}_{\nu+1} = \widetilde{A}_{\nu+1,\nu+1} x_{\nu+1}. \tag{2.3}
$$

Denoting by $P(\mathcal{R}^*, \mathcal{Q}^*)$ the canonical projection $P(\mathcal{R}^*, \mathcal{Q}^*) : \mathcal{R}^* \to \mathcal{R}^*/\mathcal{Q}^*$, let $\bar{A} : \mathcal{R}^*/\mathcal{Q}^* \to \mathcal{R}^*/\mathcal{Q}^*$ be the linear map such that $P(\mathcal{R}^*, \mathcal{Q}^*)((\widetilde{A})|\mathcal{R}^*) = \bar{A} P(\mathcal{R}^*, \mathcal{Q}^*)$. In z–coordinates, $\bar{A}$ is given exactly by $\widetilde{A}_{\nu+1,\nu+1}$.

Similarly, for each $i = 1, \dots, \nu$ consider the well–defined system

$$
\dot{x}_i = \widetilde{A}_{ii} x_i + \widetilde{B}_{ii} u_i. \tag{2.4}
$$

Since from (1.5) we have $\widetilde{B}_i \in \mathcal{R}_i^*$, it is possible to define each system (2.4) in a coordinate–free way as follows. Denoting by $P(\mathcal{R}_i^* + \mathcal{R}^*, \mathcal{R}^*)$ the canonical projection $P(\mathcal{R}_i^* + \mathcal{R}^*, \mathcal{R}^*) : \mathcal{R}_i^* + \mathcal{R}^* \to (\mathcal{R}_i^* + \mathcal{R}^*)/\mathcal{R}^*$, let $\bar{A}_i : (\mathcal{R}_i^* + \mathcal{R}^*)/\mathcal{R}^* \to (\mathcal{R}_i^* + \mathcal{R}^*)/\mathcal{R}^*$ be the linear map such that $P(\mathcal{R}_i^* + \mathcal{R}^*, \mathcal{R}^*)((\widetilde{A})|(\mathcal{R}_i^* + \mathcal{R}^*)) = \bar{A}_i P(\mathcal{R}_i^* + \mathcal{R}^*, \mathcal{R}^*)$ and let $\bar{B}_i = P(\mathcal{R}_i^* + \mathcal{R}^*, \mathcal{R}^*)(\widetilde{B}_i)$. In z–coordinates, the pair $(\bar{A}_i, \bar{B}_i)$ is exactly $(\widetilde{A}_{ii}, \widetilde{B}_{ii})$.

Finally, consider the well–defined system

$$
\dot{x}_{\nu+2} = \widetilde{A}_{\nu+2,\nu+2} x_{\nu+2} + \widetilde{B}_{\nu+2,\nu+1} u_{\nu+1}. \tag{2.5}
$$

The coordinate–free characterization of (2.5) is given by the pair $((\widetilde{A})|\mathcal{Q}^*, \Pi(\mathcal{Q}^*, \mathcal{S})(\widetilde{B}_{\nu+1}))$, where $\Pi(\mathcal{Q}^*, \mathcal{S})$ is the natural projection on $\mathcal{Q}^*$ along any $\mathcal{S}$ such that $\mathcal{Q}^* \oplus \mathcal{S} = \mathbb{R}^n$. We remark that $\Pi(\mathcal{Q}^*, \mathcal{S})(\widetilde{B}_{\nu+1})$ does not depend on the choice of $\mathcal{S}$, since $\widetilde{B}_{\nu+1} \in \mathcal{Q}^*$.

It is worth noting that, since (A, B) is controllable by assumption, also the pair $(\widetilde{A}_{ii}, \widetilde{B}_{ii})$ is controllable and, thus, (2.4) is asymptotically stabilizable via static feedback. On the other hand, also (2.5) is asymptotically stabilizable via static feedback, since $\mathcal{Q}^*$ is a controllability subspace (see proposition 1.1).

The dynamics (2.3), (2.4) and (2.5) can be *uniquely* associated to a given system (1.1), which satisfies (2.1), in the following sense. Let us apply to (2.2) any regular feedback law

$u = Fx + Gv$, which satisfies (1.5) for some column (block) partition $G = (G_1 \quad \cdots G_{\nu+1})$ and with $\mathcal{R}_i = \mathcal{R}_i^*$. Denote by $\widetilde{\Sigma}'$ the system obtained from (2.2). One might ask if the dynamics (2.3), (2.4) and (2.5), defined for (2.2), are "different" from the corresponding ones defined for $\widetilde{\Sigma}'$. Denote by (2.3)$'$, (2.4)$'$ and (2.5)$'$ the dynamics defined as (2.3), (2.4) and (2.5), respectively, on $\widetilde{\Sigma}'$. We give the following important result, which will be proved in chapter **2** for nonlinear systems.

Proposition 2.1. The dynamics (2.3) and (2.3)$'$ are equal up to changes of coordinates. The systems (2.4) and (2.4)$'$ and, respectively, (2.5) and (2.5)$'$ are equal up to changes of coordinates and regular static feedback transformations. □

Proposition 2.1 states that the definition of (2.3), (2.4) and (2.5) is *independent* (up to changes of coordinates and regular static feedback transformations) of the feedback law chosen to obtain (2.2).

2.2.2 Necessary and sufficient conditions

As already pointed out, the dynamics (2.3), (2.4) and (2.5) are crucial in finding necessary and sufficient conditions to solve NSS. Indeed, it has been shown in [90] that if NSS is solvable, then (2.3) must be asymptotically stable, i.e. $\widetilde{A}_{\nu+1,\nu+1}$ must be Hurwitz (a proof of this fact will be given for nonlinear systems in section **2**.3.2).

Conversely, assume that $\widetilde{A}_{\nu+1,\nu+1}$ is Hurwitz. Note that (2.1) plus controllability of (1.1) implies $\mathcal{R}_i^* + \mathcal{K}_i = I\!R^n$ for $i = 1, \ldots, \nu$. Indeed, if (1.2) is a regular feedback law such that (1.5) is satisfied with $\mathcal{R}_i = \mathcal{R}_i^*$ and since (A, B) is controllable, it follows from (2.1) that $\sum_{i=1}^{\nu} \mathcal{R}_j^* = I\!R^n$. This, together with $\sum_{j \neq i} \mathcal{R}_j^* \subset \mathcal{K}_i$, implies (3.1). Since (2.4) and (2.5) are asymptotically stabilizable via static feedback, it follows that, if

$$\begin{aligned} u_i &= \widetilde{F}_i x_i + v_i, \qquad i = 1, \ldots, \nu, \\ u_{\nu+1} &= \widetilde{F}_{\nu+1} x_{\nu+2} + v_{\nu+1}, \end{aligned} \tag{2.6}$$

is a feedback law such that $\widetilde{A}_{ii} + \widetilde{B}_{ii}\widetilde{F}_i$, $i = 1, \ldots, \nu$, and $\widetilde{A}_{\nu+2,\nu+2} + \widetilde{B}_{\nu+2,\nu+1}\widetilde{F}_{\nu+1}$ are Hurwitz, the system (2.2)–(2.6) is asymptotically stable and satisfies (1.4), (1.5) and (1.6) with $\mathcal{R}_i = \mathcal{R}_i^*$.

We can sum up the above results in the following theorem.

Theorem 2.2. Let us consider the class of regular static feedback laws and suppose that (A, B) is controllable. Then, NSS is solvable if and only if *a)* $\sum_{i=1}^{\nu} \mathcal{R}_i^* \cap \mathcal{B} = \mathcal{B}$,

b) the dynamics (2.3) is asymptotically stable . □

2.3 Noninteracting control with stability via dynamic feedback

As it is apparent from theorem 2.2, even for linear systems there is a nontrivial obstruction to solving NSS, given by the fact that (2.3) must be asymptotically stable. It is natural to ask if dynamic feedback can help overcoming this obstruction. To answer this question, we will show that

$$\mathcal{R}_i^* + \mathcal{K}_i = \mathbb{R}^n, \qquad i = 1, \dots, \nu, \tag{3.1}$$

is a necessary and sufficient condition to solve NSD. In other words, *noninteraction* and *stability* can be achieved via dynamic feedback if and only if the subspaces $\mathcal{R}_1^*, \dots, \mathcal{R}_\nu^*$ satisfy the output controllability property. This also amounts to say that noninteraction and stability can be achieved via dynamic feedback if and only if *only* noninteraction can be achieved via dynamic feedback. Together with theorem 2.2, this implies that if (2.3) is unstable then dynamic feedback can be used to "destroy" the unstable dynamics of (2.3). Note also that (2.1) implies (3.1).

We can assume without loss of generality that $\mathcal{Q}^* = 0$. To understand this, we will show first that NSD is solvable for (1.1) if it is solvable for the system obtained from (2.2) by projecting $\mathbb{R}^n$ onto $\mathbb{R}^n/\mathcal{Q}^*$. For, there exist a feedback law (1.2) and a column (block) partition G_1, G_2 of G such that $\mathcal{Q}^*$ is invariant under $\widetilde{A}$, $\mathcal{Q}^* \cap \mathcal{G} = \text{span}\{BG_1\}$ and $(\widetilde{A})|\mathcal{Q}^*$ is Hurwitz (this is always possible by proposition 1.1). After a change of coordinates $\eta = (\,\eta_1^T \quad \eta_2^T\,)^T$ such that $\mathcal{Q}^* = \text{span}\{\partial/\partial\eta_1\}$, the system (1.1)–(1.2) is

$$\begin{aligned} \dot\eta_1 &= \bar{A}_{11}\eta_1 + \bar{A}_{12}\eta_2 + \sum_{j=1}^{\nu+1} \bar{B}_{1j}u_j, \\ \dot\eta_2 &= \bar{A}_{22}\eta_2 + \sum_{j=1}^{\nu} \bar{B}_{2j}u_j, \\ y_i &= \bar{C}_{i2}\eta_2, \qquad i = 1, \dots, \nu, \end{aligned} \tag{3.2}$$

with $\bar{A}_{11}$ being Hurwitz. Now, consider the subsystem

$$\begin{aligned} \dot\eta_2 &= \bar{A}_{22}\eta_2 + \sum_{j=1}^{\nu} \bar{B}_{2j}u_j, \\ y_i &= \bar{C}_{i2}\eta_2, \qquad i = 1, \dots, \nu. \end{aligned} \tag{3.3}$$

If the feedback law

$$u_i = \bar{F}_{i1}\eta_2 + \bar{F}_{i2}w + \bar{G}_i v, \qquad i = 1, \dots, \nu,$$
$$\dot{w} = \bar{\Gamma}_1\eta_2 + \bar{\Gamma}_2 w + \bar{H}v,$$

solves NSD for (3.3), then by plugging it into (3.2), we obtain

$$\begin{aligned}
\dot{\eta}_1 &= \bar{A}_{11}\eta_1 + (\bar{A}_{12} + \sum_{j=1}^{\nu} \bar{B}_{1j}\bar{F}_{j1})\eta_2 + \sum_{j=1}^{\nu} \bar{B}_{1j}\bar{F}_{j2}w + \sum_{j=1}^{\nu} \bar{B}_{1j}\bar{G}_j v + \bar{B}_{1,\nu+1}u_{\nu+1},\\
\dot{\eta}_2 &= (\bar{A}_{22} + \sum_{j=1}^{\nu} \bar{B}_{2j}\bar{F}_{j1})\eta_2 + \sum_{j=1}^{\nu} \bar{B}_{2j}\bar{F}_{j2}w + \sum_{j=1}^{\nu} \bar{B}_{2j}\bar{G}_j v,\\
\dot{w} &= \bar{\Gamma}_1\eta_2 + \bar{\Gamma}_2 w + \bar{H}v,\\
y_i &= \bar{C}_{i2}\eta_2, \qquad i = 1, \dots, \nu,
\end{aligned} \tag{3.4}$$

which is asymptotically stable, since

$$\begin{pmatrix} \bar{A}_{22} + \sum_{j=1}^{\nu} \bar{B}_{2j}\bar{F}_{j1} & \sum_{j=1}^{\nu} \bar{B}_{2j}\bar{F}_{j2} \\ \bar{\Gamma}_1 & \bar{\Gamma}_2 \end{pmatrix}$$

and $\bar{A}_{11}$ are Hurwitz matrices. Moreover, (3.4) is also noninteractive by construction, since

$$\begin{aligned}
\dot{\eta}_2 &= (\bar{A}_{22} + \sum_{j=1}^{\nu} \bar{B}_{2j}\bar{F}_{1j})\eta_2 + \sum_{j=1}^{\nu} \bar{B}_{2j}\bar{F}_{2j}w + \sum_{j=1}^{\nu} \bar{B}_{2j}\bar{G}_j v,\\
\dot{w} &= \bar{\Gamma}_1\eta_2 + \bar{\Gamma}_2 w + \bar{H}v,\\
y_i &= \bar{C}_{i2}\eta_2, \qquad i = 1, \dots, \nu,
\end{aligned} \tag{3.5}$$

is noninteractive and the outputs $y_1, \dots, y_\nu$ are not influenced by (the components of) η_1. Moreover, if y_i is controllable through the inputs of the i–th block in (3.5), this remains true in (3.4) This proves that if NSD is solvable for (3.3), then it is solvable also for (3.2). Now, denote by $\bar{\mathcal{R}}_i^*$ the maximal controllability subspace for (3.3) contained in $\bigcap_{j\neq i} \ker\{\bar{C}_{j2}\}$, by $\bar{\mathcal{Q}}^*$ the maximal controllability subspace for (3.3) contained in $\bigcap_{i=1}^{\nu} \sum_{j\neq i} \bar{\mathcal{R}}_j^*$ and by $P(\mathbb{R}^n, \mathcal{Q}^*)$ the canonical projection $P(\mathbb{R}^n, \mathcal{Q}^*) : \mathbb{R}^n \to \mathbb{R}^n/\mathcal{Q}^*$. If we show that (3.3) is controllable, the distributions $\bar{\mathcal{R}}_1^*, \dots \bar{\mathcal{R}}_\nu^*$ satisfy the output controllability property and $\bar{\mathcal{Q}}^* = 0$, then, without loss of generality, we can assume $\mathcal{Q}^* = 0$ for (1.1). For, since (1.1) is controllable, (3.3) is also controllable. Since $\mathcal{Q}^* \subset \mathcal{R}_i^*$, reasoning as in [40] (lemma **A**.9), it can be easily shown that $\bar{\mathcal{R}}_i^* = P(\mathbb{R}^n, \mathcal{Q}^*)(\mathcal{R}_i^*)$ and that $\bar{\mathcal{Q}}^* = P(\mathbb{R}^n, \mathcal{Q}^*)(\mathcal{Q}^*) = 0$. By projecting (3.1), we obtain that the distributions $\bar{\mathcal{R}}_1^*, \dots \bar{\mathcal{R}}_\nu^*$ satisfy the output controllability property

Before going further, we spend few words on the lines of the proof. First, we will define a set of ν linearly independent controllability subspaces $\mathcal{R}_i^e \subset \bigcap_{j\neq i} \mathcal{K}_j^e$, $i = 1, \ldots, \nu$. Since by construction the subspaces $\mathcal{R}_1^e, \ldots, \mathcal{R}_\nu^e$ are linearly independent, there exist a regular feedback law $u^e = F^e x^e + G^e v^e$ and a column (block) partition $G_1^e, \ldots, G_{\nu+1}^e$ of G^e such that $\mathcal{R}_1^e, \ldots, \mathcal{R}_\nu^e$ are invariant under $A^e + B^e F^e$, $\mathcal{R}_i^e \cap \mathcal{G}^e = \text{span}\{B^e G_i^e\}$ and $(A^e + B^e F^e)|\mathcal{R}_i^e$ is Hurwitz. Let $\widetilde{\Sigma}^e$ be the system obtained from plugging this feedback law into (1.7). Since $\widetilde{\Sigma}^e$ is controllable, also the system, obtained from $\widetilde{\Sigma}^e$ by projecting $\mathbb{R}^{n+n^w}$ onto $\mathbb{R}^{n+n^w}/(\sum_{i=1}^{\nu} \mathcal{R}_i^e)$, is controllable. It follows that the overall system $\widetilde{\Sigma}^e$ is asymptotically stabilizable through a feedback law which preserves the invariance of the controllability subspaces $\mathcal{R}_i^e$, $i = 1, \ldots, \nu$. Since this invariance property amounts to noninteraction, we obtain both noninteraction and stability.

Now, let us go down to the details. Let s_i be the dimension of $\mathcal{R}_i^*$ and $X_{i1}, \ldots, X_{is_i}$ be linearly independent vectors such that $\mathcal{R}_i^* = \text{span}\{X_{i1} \ldots, X_{is_i}\}$.

Consider an extended system (1.7) with

$$w = \begin{pmatrix} w_1 \\ \vdots \\ w_\nu \end{pmatrix}, \; w_i \in \mathbb{R}^{s_i}, \qquad i = 1, \ldots, \nu.$$

Let $\Pi(\mathcal{R}_i^*, \mathcal{S}_i) : \mathbb{R}^n \to \mathbb{R}^{s_i}$ be a linear map such that $\Pi(\mathcal{R}_i^*, \mathcal{S}_i)(\mathcal{R}_i^*) = \mathbb{R}^{s_i}$. Given a subspace $\mathcal{S}_i$ such that $\mathcal{S}_i \oplus \mathcal{R}_i^* = \mathbb{R}^n$, a very simple choice of $\Pi(\mathcal{R}_i^*, \mathcal{S}_i)$ is given, for example, by the natural projection on $\mathcal{R}_i^*$ along $\mathcal{S}_i$. The map $\Pi(\mathcal{R}_i^*, \mathcal{S}_i)$ can be identified with a matrix $s_i \times n$ (once bases of $\mathcal{R}_i^*$ and $\mathcal{S}_i$, respectively, are chosen).

Let us define the following vectors of $\mathbb{R}^{n+n^w}$

$$X_{ik}^e = \begin{pmatrix} X_{ik} \\ 0_{s_1 \times 1} \\ \vdots \\ X_{ik}^* \\ 0_{s_{i+1} \times 1} \\ \vdots \\ 0_{s_m \times 1} \end{pmatrix}, \qquad k = 1, \ldots, s_i, \tag{3.6}$$

where

$$X_{ik}^* = \Pi(\mathcal{R}_i^*, \mathcal{S}_i)(X_{ik}),$$

and, correspondingly, the subspaces of $\mathbb{R}^{n+n^w}$

$$\mathcal{R}_i^e = \text{span}\{X_{ik}^e : k = 1, \ldots, s_i\}, \qquad i = 1, \ldots, \nu. \tag{3.7}$$

These subspaces enjoy very interesting properties.

Proposition 3.1. The subspaces $\mathcal{R}_i^e$, $i = 1, \ldots, \nu$, are linearly independent controllability subspaces for (1.7), contained in $\bigcap_{j \neq i} \mathcal{K}_j^e$ and with dimension equal to s_i; in particular, they are controlled invariant subspaces for (1.7). □

Since the subspaces $\mathcal{R}_i^e$, $i = 1, \ldots, \nu$, are independent and controlled invariant, they are also compatible.

Proposition 3.2. There exist a regular feedback law

$$u^e = F^e x^e + G^e v^e \tag{3.8}$$

and a column (block) partition $G_1^e, \ldots, G_{\nu+1}^e$ of G^e such that $(A^e + B^e F^e)\mathcal{R}_i^e \subset \mathcal{R}_i^e$ and $\mathcal{R}_i^e \cap \mathcal{B}^e = \operatorname{span}\{B^e G_i^e\}$, $i = 1, \ldots, \nu$. □

In what follows, let $v_1^e, \ldots, v_{\nu+1}^e$ be the (block) partition of the input vector v^e corresponding to the partition of G^e. Moreover, the system (1.7)–(3.8) will be denoted by

$$\begin{aligned} \dot{x}^e &= \widetilde{A}^e x^e + \sum_{j=1}^{\nu+1} \widetilde{B}_j^e v_j^e, \\ y_i &= \widetilde{C}_i^e x^e, \qquad i = 1, \ldots, \nu. \end{aligned} \tag{3.9}$$

After possibly changing coordinates, since $\mathcal{R}_1^e, \ldots, \mathcal{R}_\nu^e$ are invariant under $\widetilde{A}^e$ and $\mathcal{R}_i^e \subset \bigcap_{j \neq i} \mathcal{K}_j^e$, (3.9) has the following form

$$\begin{aligned} \dot{z}_i^e &= \widetilde{A}_{ii}^e z_i^e + \widetilde{A}_{i,\nu+1}^e z_{\nu+1}^e + \widetilde{B}_{ii}^e v_i^e + \widetilde{B}_{i,\nu+1}^e v_{\nu+1}^e, \qquad i = 1, \ldots, \nu, \\ \dot{z}_{\nu+1}^e &= \widetilde{A}_{\nu+1,\nu+1}^e z_{\nu+1}^e + \widetilde{B}_{\nu+1,\nu+1}^e v_{\nu+1}^e, \\ y_i &= \widetilde{C}_{ii}^e z_i^e + \widetilde{C}_{i,\nu+1}^e z_{\nu+1}^e, \qquad i = 1, \ldots, \nu, \end{aligned} \tag{3.10}$$

with $\mathcal{R}_i^e = \operatorname{span}\{\partial/\partial z_i^e\}$. It is worth noting that for $v_{\nu+1}^e = 0$ we obtain a system which is *noninteractive* but *not* necessarily asymptotically stable. Moreover, we have $\widetilde{C}_i^e \mathcal{R}_i^e = \mathbb{R}^{p_i}$, $i = 1, \ldots, \nu$, since $\Pi(\mathbb{R}^{n+n^w}, \mathbb{R}^n)(\mathcal{R}_i^e) = \mathcal{R}_i^*$, where $\Pi(\mathbb{R}^{n+n^w}, \mathbb{R}^n) : \mathbb{R}^{n+n^w} \to \mathbb{R}^n$ is the natural projection, and $\mathcal{R}_i^* + \mathcal{K}_i = \mathbb{R}^n$.

We will see in a moment that it is possible to stabilize (3.9) without destroying its output controllability and noninteraction properties.

For, note that, since $\mathcal{R}_i^e = \langle \widetilde{A}^e | \operatorname{span}\{\widetilde{B}_i^e\} \rangle$, the pair $(\widetilde{A}_{ii}^e, \widetilde{B}_{ii}^e)$ is controllable. Thus, each system

$$\dot{z}_i^e = \widetilde{A}_{ii}^e z_i^e + \widetilde{B}_{ii}^e v_i^e, \tag{3.11}$$

is asymptotically stabilizable through a feedback law

$$v_i^e = \widetilde{F}_i^e z_i^e + v_i. \tag{3.12}$$

On the other hand, also

$$\dot{z}_{\nu+1}^e = \widetilde{A}_{\nu+1,\nu+1}^e z_{\nu+1}^e + \widetilde{B}_{\nu+1,\nu+1}^e v_{\nu+1}^e \tag{3.13}$$

is asymptotically stabilizable through a feedback law

$$v_{\nu+1}^e = \widetilde{F}_{\nu+1}^e z_{\nu+1}^e, \tag{3.14}$$

since (3.10) is controllable and (3.13) is the system obtained from (3.10) by projecting $\mathbb{R}^{n+n^w}$ onto $\mathbb{R}^{n+n^w}/(\sum_{i=1}^{\nu} \mathcal{R}_i^e)$.

The system (3.10)–(3.12)–(3.14) is clearly asymptotically stable (since $\widetilde{A}_{ii}^e + \widetilde{B}_{ii}^e \widetilde{F}_i^e$, $i = 1, \ldots, \nu+1$, are Hurwitz) and noninteractive (since (3.12) and (3.14) preserve the triangular structure of (3.10)).

At this point, a problem may arise when $\sum_{i=1}^{\nu} \dim v_i > m$. In this case, by applying the celebrated Heymann's lemma ([46]), the feedback law $u^e = F^e x^e + G^e v^e$ of proposition 3.2 can be chosen in such a way that, if $\widetilde{b}_i^e$ is a nonzero column of $\widetilde{B}_i^e$, we have $\mathcal{R}_i^e = \langle \widetilde{A}^e | \mathrm{span}\{\widetilde{b}_i^e\} \rangle$. It is worth noting that, by doing so, the number of the inputs is reduced to ν. If $\nu < m$, by cascading (3.8), (3.12) and (3.14), we obtain (in (x, w)–coordinates) a dynamic feedback law

$$\begin{aligned} u &= F_1 x + F_2 w + G v, \\ \dot{w} &= \Gamma_1 x + \Gamma_2 w + H v, \end{aligned} \tag{3.15}$$

such that the system

$$\begin{aligned} \dot{x} &= Ax + B(F_1 x + F_2 w) + BGv, \\ \dot{w} &= \Gamma_1 x + \Gamma_2 w + Hv, \\ u &= F_1 x + F_2 w + Gv, \end{aligned}$$

with output vector u and input vector v is possibly *noninvertible*, in the sense of [74] (see also section **2**.1). Equivalently, this can be restated in the Laplace domain by saying that we obtain a proper compensator $G(s)$ with certain rows linearly dependent for all s (or identically equal to zero).

So far, we have proved that if (3.1) is satisfied then NSD is solvable. The converse of this result follows easily from $\mathcal{R}_i^e + \mathcal{K}_i^e = \mathbb{R}^{n+n^w}$, $i = 1, \ldots, \nu$, and the following proposition.

Proposition 3.3. ([91], lemma 9.1) *Let* $\Pi(\mathbb{R}^{n+n^w}, \mathbb{R}^n)$ *be as above. Then,* $\Pi(\mathbb{R}^{n+n^w}, \mathbb{R}^n)(\mathcal{R}_i^e)$ *is a controllability subspace for (1.1) contained in* $\bigcap_{j\neq i} \mathcal{K}_j$. *In particular,* $\Pi(\mathbb{R}^{n+n^w}, \mathbb{R}^n)(\mathcal{R}_i^e) \subset \mathcal{R}_i^*$. □

We sum up the above results in the following important theorem.

Theorem 3.4. ([58], [90], [91]) *Assume that* (A, B) *is controllable. Then, NSD is solvable if and only if*

$$\mathcal{R}_i^* + \mathcal{K}_i = \mathbb{R}^n, \qquad i = 1, \ldots, \nu.$$ □

Before concluding this section, we want to spend few more words on the output controllability condition (3.1). If $m = p = \nu$, i.e. each output is controlled by one input, the condition (3.1) guarantees that $\widetilde{\Sigma}^e$ has vector relative degree $r_1^e, \ldots, r_m^e$. Indeed, if $\mathcal{R}_i^e$ is defined as in (3.7) and a feedback law, which solves ND for (1.1), is chosen as above, we have $\mathcal{R}_i^e = \langle \widetilde{A}^e | \mathrm{span}\{\widetilde{b}_i^e\}\rangle$ and $\mathcal{R}_i^e + \mathcal{K}_i^e = \mathbb{R}^{n+n^w}$ for $i = 1, \ldots, \nu$. This implies that $C_i^e(\widetilde{A}^e)^{r_i^e}\widetilde{b}_i^e \neq 0$ for $i = 1, \ldots, \nu$ and for some $r_i^e < s_i$ or, equivalently, that $\widetilde{\Sigma}^e$ has vector relative degree $r_1^e, \ldots, r_m^e$.

On the other hand, if $\widetilde{\Sigma}^e$ has vector relative degree $r_1^e, \ldots, r_m^e$, then $\mathcal{R}_i^e + \mathcal{K}_i^e = \mathbb{R}^{n+n^w}$, $i = 1, \ldots, m$, where $\mathcal{R}_i^e$ is defined as in (1.5). By projecting $\mathcal{R}_i^e + \mathcal{K}_i^e = \mathbb{R}^{n+n^w}$, we obtain $\Pi(\mathbb{R}^{n+n^w}, \mathbb{R}^n)(\mathcal{R}_i^e) + \mathcal{K}_i = \mathbb{R}^n$. This, together with proposition 3.3, gives (3.1). Thus, the output controllability condition (3.1) is equivalent to requiring $\widetilde{\Sigma}^e$ have vector relative degree $r_1^e, \ldots, r_m^e$. For nonlinear systems, this equivalence is no longer true, in general. The reasons of this fact will be explained in more details in the next chapter.

CHAPTER 3

NONINTERACTING CONTROL WITH LOCAL STABILITY FOR NONLINEAR SYSTEMS: BASIC PRINCIPLES

3.1 The class of admissible feedback laws

Let us consider nonlinear systems of the form

$$\begin{aligned} \dot{x} &= f(x) + \sum_{j=1}^{m} g_j(x) u_j, \\ y_i &= h_i(x), \qquad i = 1, \ldots, m, \end{aligned} \tag{1.1}$$

with $x \in \mathbb{R}^n$, $u_j \in \mathbb{R}$, $f(x)$ and $g_j(x)$, $j = 1, \ldots, m$, smooth vector fields and $h_i(x)$, $i = 1, \ldots, m$, smooth real-valued functions, defined on some open set of $\mathbb{R}^n$, and dynamic feedback laws of the form

$$\begin{aligned} u_i &= \alpha_i(x, w) + \sum_{j=1}^{m} \beta_{ij}(x, w) v_j, \qquad i = 1, \ldots, m, \\ \dot{w} &= \gamma(x, w) + \sum_{j=1}^{m} \delta_j(x, w) v_j, \end{aligned} \tag{1.2}$$

with $w \in \mathbb{R}^{n^w}$, $v_j \in \mathbb{R}$ and $\alpha_i(x,w), \beta_{ij}(x,w), \gamma(x,w)$ and $\delta_j(x,w)$ smooth (vector-valued) functions, defined on some open set of $\mathbb{R}^{n+n^w}$. Moreover, $\alpha_i(0,0) = 0$ and $\gamma(0,0) = 0$. If $n^w = 0$, the feedback law is said to be *static*.

As well-known in the literature, for linear systems (2.1.1) with $m = p = \nu$, the design of a linear feedback law which achieves noninteraction plus output controllability amounts to find a proper $m \times m$ matrix $G(s)$ such that, denoting by $W(s)$ the transfer matrix associated with (2.1.1), the matrix $W(s)G(s)$ is diagonal, with diagonal entries $d_i(s)$, $i = 1, \ldots, m$, of the form $1/s^l$ for some integer l. This implies that $G(s)$ and $W(s)$ be invertible matrices for almost all s. We will say that a linear system (2.1.1) with $m = p = \nu$ is *invertible* if its transfer function matrix is invertible for almost all s.

On the other hand, the existence of a proper $G(s)$, which is invertible for almost all s, amounts to the existence of a feedback law

$$u_i = F_{i1}x + F_{i2}w + \sum_{j=1}^{m} G_{ij}v_j, \qquad i = 1, \ldots, m,$$
$$\dot{w} = \Gamma_1 x + \Gamma_2 w + \sum_{j=1}^{m} H_j v_j,$$

such that the linear system

$$\begin{aligned} u_i &= F_{i1}x + F_{i2}w + \sum_{j=1}^{m} G_{ij}v_j, \qquad i = 1, \ldots, m, \\ \dot{w} &= \Gamma_1 x + \Gamma_2 w + \sum_{j=1}^{m} H_j v_j, \\ \dot{x} &= Ax + \sum_{j=1}^{m} B_j(F_{j1}x + F_{j2}w) + \sum_{j=1}^{m}\sum_{h=1}^{m} B_j G_{jh} v_h, \end{aligned} \tag{1.3}$$

with input vector $v = (\, v_1 \quad \cdots \quad v_m \,)^T$ and output vector $u = (\, u_1 \quad \cdots \quad u_m \,)^T$ is invertible in the sense of [74]. We will say in this case that the feedback law is *invertible*.

A very easy way of testing the invertibility of (1.3) (and, in general, of a linear system) is given by the *Silverman's structure algorithm* ([74]). This algorithm gives at the k–th step a suitable matrix D_k. If there exists $k^* \leq n$ such that the rank of D_{k^*} is m, then (1.3) is invertible. Conversely, if (1.3) invertible, then there must exist some $k^* \leq n$ such that D_{k^*} has rank equal to m.

At the beginning of the eighties, the concept of invertibility has been extended to a nonlinear setting. On the base of a first algorithm given by Hirschorn ([48]), Singh ([77]) has given a generalized version of the Silverman's structure algorithm for the class of systems

$$\begin{aligned} \dot{x} &= f(x) + \sum_{j=1}^{m} g_j(x)u_j, \\ y_i &= h_i(x) + \sum_{j=1}^{m} k_{ij}(x)u_j, \qquad i = 1, \ldots, m. \end{aligned} \tag{1.4}$$

Since Singh's algorithm plays a fundamental role in all the subsequent analysis, we will shortly review it here. We will say that (1.4) is *invertible* at $\widehat{x} \in \mathbb{R}^n$ if, denoting by $y(t, \widehat{x}, u(t))$ the output of (1.4) with initial state $\widehat{x}$ and input vector $u(t)$, whenever $u^a(t)$ and $u^b(t)$ are distinct admissible input vectors, we have $y(t, u^a(t), \widehat{x}) \neq y(t, u^b(t), \widehat{x})$. The system (1.4) is *(strongly) invertible* if, for each $\hat{x} \in M$, with M open and dense subset of $\mathbb{R}^n$, there exists an open neighbourhood U of $\hat{x}$ such that (1.4) is invertible at each $x \in U$. Invertibility at $\hat{x}$ does not imply, in general, invertibility at any other point $x \neq \hat{x}$ (this is due to the nonlinear dependence of $y(t, u(t), \hat{x})$ on the initial state $\hat{x}$). On the other hand, for linear systems this definition of invertibility is equivalent to the one given above in the Laplace domain and does not depend on the initial state $\hat{x}$.

Let us denote $(v_1 \ \cdots \ v_m)^T$ by v, $(u_1 \ \cdots \ u_m)^T$ by u, $(y_1 \ \ldots \ y_m)^T$ by y, $(h_1(x) \ \ldots \ h_m(x))^T$ by $h(x)$, $(g_1(x) \ \ldots \ g_m(x))$ by $G(x)$, $(k_{i1}(x) \ \cdots \ k_{im}(x))$ by $K_i(x)$ and $(K_1^T(x) \ \cdots \ K_m^T(x))^T$ by $K(x)$. Given a vector-valued function $Y(x)$ and a matrix $A(x)$ (with suitable dimensions), we also denote by $L_A Y(x)$ the matrix $(\partial Y/\partial x)(x)A(x)$, where $(\partial Y/\partial x)(x)$ is the Jacobian of $Y(x)$ at x.

Suppose that the rank of $K(x)$ is q_0 for all x in an open and dense subset M_0 of $\mathbb{R}^n$. After possibly permuting and linearly combining the rows of $K(x)$, we obtain

$$\dot{x} = f(x) + G(x)u, \qquad x \in M_0,$$
$$z_1(x, y) = c_1(x) + D_1(x)u,$$

where

$$D_1(x) = R_0(x)K(x) = \begin{pmatrix} D_{11}(x) \\ 0 \end{pmatrix},$$
$$z_1(x, y) = R_0(x)y, \ c_1(x) = R_0(x)h(x),$$

$R_0(x)$ is a $m \times m$ matrix, with smooth entries and invertible for all $x \in M_0$, and $D_{11}(x)$ is a $q_0 \times m$ matrix with rank q_0 for all $x \in M_0$. Now, decompose z_1 as follows

$$z_1(x, y) = \begin{pmatrix} \bar{z}_1(x, y) \\ \hat{z}_1(x, y) \end{pmatrix} = \begin{pmatrix} \bar{c}_1(x) \\ \hat{c}_1(x) \end{pmatrix} + \begin{pmatrix} D_{11}(x) \\ 0 \end{pmatrix} u.$$

In what follows, we will denote by the superscript (k) the number of derivatives with respect to time.

Step 1. Let us differentiate $\hat{z}_1(x, y)$ and $\hat{c}_1(x)$ with respect to time. We obtain

$$\hat{z}_1^{(1)}(x, y, y^{(1)}, u) = L_f \hat{z}_1(x, y) + \frac{\partial \hat{z}_1}{\partial y}(x, y) y^{(1)} + L_G \hat{z}_1(x, y)u,$$
$$\hat{c}_1^{(1)}(x, u) = L_f \hat{c}_1(x) + L_G \hat{c}_1(x)u,$$

Assume that the rank of the matrix

$$A_1(x, y) = \begin{pmatrix} D_{11}(x) \\ L_G \hat{c}_1(x) - L_G \hat{z}_1(x, y) \end{pmatrix}$$

is equal to $q_1 \geq q_0$ on a dense and open subset $M_1 \subset M_0 \times I\!R^m$. After possibly permuting and linearly combining the rows of $A_1(x,y)$, since $\widehat{z}_1^{(1)}(x,y,y^{(1)},u) = \widehat{c}_1^{(1)}(x,u)$, we obtain

$$\begin{aligned} \dot{x} &= f(x) + G(x)u, \qquad (x,y) \in M_1, \\ z_2(x,y,y^{(1)}) &= c_2(x,y) + D_2(x,y)u, \end{aligned}$$

where

$$\begin{aligned} D_2(x,y) &= R_1(x,y)A_1(x,y) = \begin{pmatrix} D_{22}(x,y) \\ 0 \end{pmatrix}, \\ z_2(x,y,y^{(1)}) &= R_1(x,y)\begin{bmatrix} \bar{z}_1(x) \\ L_f\widehat{z}_1(x,y) + \frac{\partial \widehat{z}_1}{\partial y}(x,y)y^{(1)} \end{bmatrix}, \\ c_2(x,y) &= R_1(x,y)\begin{pmatrix} \bar{c}_1(x) \\ L_f\widehat{c}_1(x) \end{pmatrix}, \end{aligned}$$

$R_1(x,y)$ is a $m \times m$ matrix, with smooth entries and invertible for all $(x,y) \in M_1$, and $D_{22}(x,y)$ is a $q_1 \times m$ matrix with rank q_1 for all $(x,y) \in M_1$. Now decompose z_2 as follows

$$z_2(x,y,y^{(1)}) = \begin{pmatrix} \bar{z}_2(x,y,y^{(1)}) \\ \widehat{z}_2(x,y,y^{(1)}) \end{pmatrix} = \begin{pmatrix} \bar{c}_2(x,y) \\ \widehat{c}_2(x,y) \end{pmatrix} + \begin{pmatrix} D_{22}(x,y) \\ 0 \end{pmatrix} u.$$

Step $k \geq 2$. Let us differentiate $\widehat{z}_k(x,y,\ldots,y^{(k-1)})$ and $\widehat{c}_k(x,y,\ldots,y^{(k-2)})$ with respect to time. We obtain

$$\begin{aligned} \widehat{z}_k^{(1)}(x,y,\ldots,y^{(k)},u) &= L_f\widehat{z}_k(x,y,\ldots,y^{(k-1)}) + \sum_{j=0}^{k-1} \frac{\partial \widehat{z}_k}{\partial y^{(j)}}(x,y,\ldots,y^{(k-1)})y^{(j+1)} + \\ &\quad + L_G\widehat{z}_k(x,y,\ldots,y^{(k-1)})u, \\ \widehat{c}_k^{(1)}(x,y,\ldots,y^{(k-1)},u) &= L_f\widehat{c}_k(x,y,\ldots,y^{(k-2)}) + L_G\widehat{c}_k(x,y,\ldots,y^{(k-2)})u + \\ &\quad + \sum_{j=0}^{k-2} \frac{\partial \widehat{c}_k}{\partial y^{(j)}}(x,y,\ldots,y^{(k-2)})y^{(j+1)}. \end{aligned}$$

Suppose that the rank of the matrix

$$A_k(x,y,\ldots,y^{(k-1)}) = \begin{pmatrix} D_{kk}(x,y,\ldots,y^{(k-2)}) \\ L_G\widehat{c}_k(x,y,\ldots,y^{(k-2)}) - L_G\widehat{z}_k(x,y,\ldots,y^{(k-1)}) \end{pmatrix}$$

is equal to $q_k \geq q_{k-1}$ on a dense and open subset $M_k \subset M_{k-1} \times I\!R^m$. After possibly permuting and linearly combining the rows of $A_k(x,y,\ldots,y^{(k-1)})$, since $\widehat{z}_k^{(1)}(x,y,\ldots,y^{(k)},u) = \widehat{c}_k^{(1)}(x,y,\ldots,y^{(k-1)},u)$, we obtain

$$\begin{aligned} \dot{x} &= f(x) + G(x)u, \qquad (x,y,\ldots,y^{(k-1)}) \in M_k, \\ z_{k+1}(x,y,\ldots,y^{(k)}) &= c_{k+1}(x,y,\ldots,y^{(k-1)}) + D_{k+1}(x,y,\ldots,y^{(k-1)})u, \end{aligned}$$

where

$$D_{k+1}(x,y,\ldots,y^{(k-1)}) = R_k(x,y,\ldots,y^{(k-1)})A_k(x,y,\ldots,y^{(k-1)}) = \\ = \begin{pmatrix} D_{k+1,k+1}(x,y,\ldots,y^{(k-1)}) \\ 0 \end{pmatrix},$$

$$z_{k+1}(x,y,\ldots,y^{(k)}) = R_k(x,y,\ldots,y^{(k-1)}) \cdot \\ \begin{pmatrix} \bar{z}_k(x,y,\ldots,y^{(k-2)}) \\ L_f\widehat{z}_k(x,y,\ldots,y^{(k-1)}) + \sum_{j=0}^{k-1} \frac{\partial \widehat{z}_k}{\partial y^{(j)}}(x,y,\ldots,y^{(k-1)})y^{(j+1)} \end{pmatrix},$$

$$c_{k+1}(x,y,\ldots,y^{(k-1)}) = R_k(x,y,\ldots,y^{(k-1)}) \cdot \\ \begin{pmatrix} \bar{c}_k(x,y,\ldots,y^{(k-2)}) \\ L_f\widehat{c}_k(x,y,\ldots,y^{(k-2)}) + \sum_{j=0}^{k-2} \frac{\partial \widehat{c}_k}{\partial y^{(j)}}(x,y,\ldots,y^{(k-2)})y^{(j+1)} \end{pmatrix},$$

$R_k(x,y,\ldots,y^{(k-1)})$ is a $m \times m$ matrix, with smooth entries and invertible for all $(x,y,\ldots,y^{(k-1)}) \in M_k$, and $D_{k+1,k+1}(x,y,\ldots,y^{(k-1)})$ is a $q_k \times m$ matrix with rank q_k for all $(x,y,\ldots,y^{(k-1)}) \in M_k$. Now decompose z_{k+1} as follows

$$z_{k+1}(x,y,\ldots,y^{(k)}) = \begin{pmatrix} \bar{z}_{k+1}(x,y,\ldots,y^{(k)}) \\ \widehat{z}_{k+1}(x,y,\ldots,y^{(k)}) \end{pmatrix} = \begin{pmatrix} \bar{c}_{k+1}(x,y,\ldots,y^{(k-1)}) \\ \widehat{c}_{k+1}(x,y,\ldots,y^{(k-1)}) \end{pmatrix} + \\ + \begin{pmatrix} D_{k+1,k+1}(x,y,\ldots,y^{(k-1)}) \\ 0 \end{pmatrix} u.$$

Theorem 1.1. ([77]) *If there exists $k^* < \infty$ such that $q_{k^*} = m$ on an open and dense subset $M_{k^*} \subset \mathbb{R}^n \times \mathbb{R}^{mk^*}$, then (1.3) is invertible as long as its trajectory satisfies $(x(t),y(t),\ldots,y^{(k^*-1)}(t)) \in M_{k^*}$.* □

If $f(x)$, $g_j(x)$ and $h_j(x)$, $i = 1,\ldots,m$, are analytic functions of x, then for each k the rank of the matrix $A_k(x,y,\ldots,y^{(k-1)})$ is always constant on some open and dense subset $M_k \subset \mathbb{R}^n \times \mathbb{R}^{mk}$.

The following definition extends to a nonlinear setting the concept of invertible feedback law. Let $\beta(x,w)$ be the $m \times m$ matrix with $\beta_{ij}(x,w)$ as (i,j)–th entry and $\alpha(x,w)$ the $m \times 1$ vector with $\alpha_i(x,w)$ as i–th entry. Moreover, let $\delta(x,w)$ be the $n^w \times m$ matrix with $\delta_i(x,w)$ as i–th column.

Definition 1.2. A feedback law (1.2) is said to be invertible if the system

$$\begin{aligned} u &= \alpha(x,w) + \beta(x,w)v, \\ \dot{x} &= f(x) + G(x)\alpha(x,w) + G(x)\beta(x,w)v, \\ \dot{w} &= \gamma(x,w) + \delta(x,w)v, \end{aligned} \tag{1.5}$$

with input vector v and output vector u is invertible. □

If $n^w = 0$ and $\beta(x)$ is invertible for all x in an open and dense subset of $\mathbb{R}^n$, (1.2) is invertible (in this case, we will say that (1.2) is *regular*; on the other hand, an invertible static feedback law is *not* necessarily regular at all points of an open and dense subset of $\mathbb{R}^n$). Similarly, if $n^w > 0$ and $\beta(x, w)$ is invertible for all (x, w) in an open and dense subset of $\mathbb{R}^{n+n^w}$, (1.2) is invertible.

Another interesting case in which (1.2) is invertible is when (1.5) has vector relative degree at each point of an open and dense subset of $\mathbb{R}^{n+n^w}$: in this case, if $r_1, \ldots, r_m$ is the vector relative degree of (1.5) at (x, w), Singh's algorithm stops after $k^* = (\max_{i \in \{1,\ldots,m\}} r_i)$-th iterations and the matrix A_{k^*} is exactly the decoupling matrix of (1.5).

Suppose now that the systems (1.1) and

$$\begin{aligned} \dot{x} &= f(x) + G(x)\alpha(x, w) + G(x)\beta(x, w)v, \\ \dot{w} &= \gamma(x, w) + \delta(x, w)v, \\ y_i &= h_i(x), \qquad i = 1, \ldots, m, \end{aligned} \tag{1.6}$$

are both noninteractive and have vector relative degree at each point of an open and dense subset $M \subset \mathbb{R}^n$ and $M \times M^w \subset \mathbb{R}^{n+n^w}$, respectively. It is possible to show that the feedback law (1.2) is invertible. This is a straightforward consequence of the following proposition, which is a slight modification of a result due to Wagner (in [86] and [87], the feedback laws (1.2) such that (1.6) has vector relative degree at all points of an open set, containing the origin of the state space, are said to be *regular*).

Proposition 1.3. Assume that both (1.1) and (1.6) are noninteractive and have vector relative degree at each point of an open and dense subset $M \subset \mathbb{R}^n$ and $M \times M^w \subset \mathbb{R}^{n+n^w}$, respectively. Then, $\beta_{ij}(x, w) = 0$ for $j \neq i$ and for all $(x, w) \in M \times M^w$. Moreover, at each $(x^\circ, w^\circ) \in M \times M^w$ either one of the following facts can happen:

a) $\beta_{ii}(x, w) \neq 0$ for all (x, w) in a neighbourhood of x°, w°;

b) $\beta_{ii}(x, w) = 0$ for all (x, w) in an open neighbourhood of (x°, w°); in addition, there exists $\varrho_i \geq 1$ such that $L_{g_j} L_f^k \alpha_i(x, w) = 0$ for $0 \leq k < \varrho_i - 1$, $1 \leq j \leq m$ and for all (x, w) in an open neighbourhood of (x°, w°) and $L_{g_i} L_f^{\varrho_i - 1} \alpha_i(x, w) \neq 0$ for all (x, w) in a neighbourhood of x°, w°.

In other words, the system (1.5) (with output vector u) has vector relative degree at each point (x°, w°) of an open and dense subset of $\mathbb{R}^{n+n^w}$. □

Now, let

$$\widetilde{f}^e(x,w) = \begin{pmatrix} f(x) + G(x)\alpha(x,w) \\ \gamma(x,w) \end{pmatrix}, \quad \widetilde{g}^e_j(x,w) = \begin{pmatrix} G(x)\beta_j(x,w) \\ \delta_j(x,w) \end{pmatrix}, \quad j = 1,\ldots,m.$$

Definition 1.4. An invertible feedback law is said to be weakly noninteractive if for all (x,w) *in an open subset of* $\mathbb{R}^{n+n^w}$

$$\begin{aligned} \beta_{ij}(x,w) &= 0, & j &\neq i, \\ L_{\widetilde{g}^e_j}\widetilde{D}^e\alpha_i(x,w) &= 0, & j &\neq i, \\ L_{\widetilde{g}^e_j}\widetilde{D}^e\beta_{ii}(x,w) &= 0, & j &\neq i, \end{aligned} \tag{1.7}$$

where $\widetilde{D}^e$ *is an arbitrary composition of the Lie derivatives* $L_{\widetilde{f}^e}$ *and* $L_{\widetilde{g}^e_j}$, $j = 1,\ldots,m$. □

In [86] and [87] the feedback laws, which satisfy (1.7), are called *noninteraction* feedback laws, but here we prefer the more precise terminology given above. Indeed, condition (1.7) is necessary (but *not* sufficient, in general) for (1.5) to be noninteractive (with output vector u). It is also sufficient if we assume, for example, that $f(x)$, $g_i(x)$ and $h_i(x)$, $i = 1,\ldots,m$, are analytic functions of x (see [55]).

Weakly noninteractive feedback laws will arise naturally in solving the problem of achieving noninteraction and stability for some significant classes of systems. Note that the class of invertible feedback laws contains the class of weakly noninteractive feedback laws. The following proposition states that, under suitable assumptions, these two classes do coincide (the proof can be found in [86]).

Proposition 1.5. Under the assumptions of proposition 1.3, any invertible feedback law is weakly noninteractive. □

Throughout this notes, we will consider only invertible feedback laws and assume that for these feedback laws the subset M_{k^*} contains the origin.

3.2 Problem formulation

In this section, we will formulate the problem of achieving noninteraction and stability for nonlinear systems (1.1). This formulation is slightly different from the one given in section 2.2, in the sense that here we require the closed–loop system (1.6) to be noninteractive and to have vector relative degree at the origin of the state space. In the case of linear systems, the two formulations are equivalent (see the end of section **2**.4). On the other hand, for nonlinear systems this equivalence is no longer true, as it will be shown in a moment. A formulation, which follows the one given in section **2**.2, will be given in the case of nonlinear systems with block-partitioned outputs, for which the concept of relative degree makes no more sense (see section **4**.4.3).

A necessary condition for (1.1) to be noninteractive is

$$\begin{aligned} L_{g_i}h_i(x) &= 0, \\ L_{g_j}L_{g_{j_s}}\cdots L_{g_{j_1}}h_i(x) &= 0, \qquad j \neq i,\ s \geq 1,\ j_s \in \{0,\ldots,m\}, \end{aligned} \tag{2.1}$$

for all x in an open neighbourhood of x_0 and with $g_0(x) = f(x)$. If we take $j = i$ in (2.1), we obtain a necessary condition for the i-th output not to be influenced by the i-th input. As a consequence, if at least one of the terms $L_{g_i}h_i(x_0)$ or $L_{g_i}L_{g_{j_s}}\cdots L_{g_{j_1}}h_i(x_0)$ is nonzero, then the i-th input *does* influence the i-th output. This is, for example, a characteristic property of noninteractive nonlinear systems (1.1) which have vector relative degree $r_1,\ldots,r_m$ at x_0. Indeed, in this case the terms $L_{g_i}L_f^{r_i-1}h_i(x_0)$, $i = 1,\ldots,m$, are nonzero.

Motivated by the above facts, the problem of achieving noninteraction and stability can be formulated as follows, according to the class of feedback laws considered (this feedback laws are defined in a neighborhood of the origin of the state space: globally defined feedback laws will be considered in chapter **3**).

Noninteracting control problem with local stability via static feedback laws (NLSS). *Find, if possible, a static feedback law (1.2), defined in a neighborhhood of* x_0*, such that the system (1.1)-(1.2) is noninteractive, locally asymptotically stable in* x_0 *and has vector relative degree at* x_0.

Noninteracting control problem with local stability via dynamic feedback laws (NLSD). *Find, if possible,* $n^w \geq 0$ *and a dynamic feedback law (1.2), defined in a neighbourhood of* $(x_0, w_0) = (0,0)$*, such that the system (1.1)-(1.2) is noninteractive, locally asymptotically stable in* (x_0, w_0) *and has vector relative degree at* (x_0, w_0).

Let us spend few words on comparing this formulation with the one given for linear systems. Let $\mathcal{K}_i$ be the smooth distribution which assign at each x the subspace of $\mathbb{R}^n$ spanned by the vectors d_i such that $\langle dh_i(x), d_i\rangle = 0$ and $\mathcal{R}_i^*$ be the maximal local controllability distribution for (1.1) contained in $\bigcap_{j\neq i}\mathcal{K}_j$ (supposed to exist). Moreover, for a given feedback law $u = \alpha(x) + \beta(x)v$, let $\tilde{f}(x) = (f + G\alpha)(x)$, $\tilde{g}_j(x)$ be the j–th column of $(G\beta)(x)$ and $ad^k_{\tilde{f}}\tilde{g}_i$, $k \geq 0$, be the smooth vector fields defined recursively as

$$ad^0_{\tilde{f}}\tilde{g}_i(x) = \tilde{g}_i(x),$$
$$ad^k_{\tilde{f}}\tilde{g}_i(x) = [\tilde{f}, ad^{k-1}_{\tilde{f}}\tilde{g}_i](x), \qquad k \geq 0,$$

where $[\cdot,\cdot]$ denotes the Lie bracket of two vector fields.

Assume that a given feedback law $u = \alpha(x) + \beta(x)v$ solves NLSS and let $\tilde{\Sigma}$ be the noninteractive closed–loop system. As it has been shown at the end of section **2.3**, the condition $\mathcal{R}^*_i + \mathcal{K}_i = \mathbb{R}^n$, $i = 1, \dots, m$, is equivalent, in the case of linear systems, to the property that $\tilde{\Sigma}$ has vector relative degree. Similarly, for nonlinear systems (1.1), if $\tilde{\Sigma}$ has vector relative degree $r_1, \dots, r_m$ at x_0, then $\mathcal{R}^*_i + \mathcal{K}_i = \mathbb{R}^n$ in a neighbourhood of x_0. Indeed, since $\tilde{\Sigma}$ has vector relative degree $r_1, \dots, r_m$ at x_0, it can be easily shown that $L_{ad^k_{\tilde{f}}\tilde{g}_i} h_i(x) = 0$ for $k < r_i - 1$ and for all x in a neighbourhood of x_0 and

$$L_{ad^{r_i-1}_{\tilde{f}}\tilde{g}_i} h_i(x) \neq 0 \tag{2.2}$$

for all x in a neighbourhood of x_0. Since

$$\text{span}\{ad^k_{\tilde{f}}\tilde{g}_i : k \geq 0\} \subset \mathcal{R}^*_i, \qquad i = 1, \dots, m, \tag{2.3}$$

(see proposition 3.3.1), (2.2) implies that $\mathcal{R}^*_i + \mathcal{K}_i = \mathbb{R}^n$ in a neighbourhood of x_0. The converse of this result is *not* true in general. Indeed, since the inclusion (2.3) may happen to be strict (while, for linear systems, it holds *always* with equality), we could as well have $L_{ad^k_{\tilde{f}}\tilde{g}_i} h_i(x) = 0$ for $k \geq 0$ and for all x in a neighbourhood of x_0 but $L_{X_i} h_i(x) \neq 0$ for some $X_i \in \mathcal{R}^*_i$ and for all x in a neighbourhood of x_0. Moreover, even if $\text{span}\{ad^k_{\tilde{f}}\tilde{g}_i : k \geq 0\} + \mathcal{K}_i = \mathbb{R}^n$, not necessarily $\tilde{\Sigma}$ has vector relative degree at x_0, as it results, for example, for the noninteractive system

$$\begin{aligned}
\dot{x}_1 &= x_2, \\
\dot{x}_2 &= u_1,. \\
\dot{x}_3 &= u_2, \\
y_1 &= x_1 + x_2^2, \\
y_2 &= x_3.
\end{aligned}$$

The above reasoning can be repeated for the case of dynamic feedback. Let $\mathcal{K}^e_i$ be the distribution which assigns at each point (x, w) the subspace $\mathcal{K}_i(x) \oplus \mathbb{R}^{n^w}$, where $\oplus$ denotes the external direct sum of vector spaces. Moreover, with $\widetilde{f}^e, \widetilde{g}_1^e, \dots, \widetilde{g}_m^e$ as in section **2**.1, let $\mathcal{R}_i^e$ be the smallest distribution which is locally invariant under $\widetilde{f}^e, \widetilde{g}_1^e, \dots, \widetilde{g}_m^e$ and contains span $\{\widetilde{g}_i^e\}$. Moreover, let (1.2) be a feedback law which solves NLSD. If (1.1)–(1.2) has vector relative degree at (x_0, w_0), then $\mathcal{R}_i^e + \mathcal{K}_i^e = \mathbb{R}^{n+n^w}$, $i = 1 \dots m$, for all (x, w) in a neighbourhood of (x_0, w_0). Since invertible laws are considered, as it will be shown in section **2**.4, the distribution $\mathcal{R}_i^e$ projects down to $\mathcal{R}_i^*$, i.e. $\pi_{*(x,w)}(\mathcal{R}_i^e)(x, w) = \mathcal{R}_i^*(\pi(x, w))$, where π is the natural projection $\pi : \mathbb{R}^{n+n^w} \to \mathbb{R}^n$ and $\pi_{*(x,w)}$ is the Jacobian of π at (x, w). As a consequence, by projecting $\mathcal{R}_i^e + \mathcal{K}_i^e = \mathbb{R}^{n+n^w}$, we obtain $\mathcal{R}_i^* + \mathcal{K}_i = \mathbb{R}^n$. In other words, if $\widetilde{\Sigma}^e$ has vector relative degree at (x_0, w_0), then $\mathcal{R}_i^* + \mathcal{K}_i = \mathbb{R}^n$, $i = 1, \dots, m$. The converse is *not* true in general, as it can be shown by reasoning as in the case of static feedback.

In what follows, we will use the notations NS and ND to denote, respectively, the problem of noninteracting control via static feedback (without the requirement of internal stability) and the problem of noninteracting control via dynamic feedback (without the requirement of internal stability). Moreover, we consider only state–feedback laws. Before ending this section, we give some basic notations, which will be freely used throughout this chapter. Let

$$G(x) = (\, g_1(x) \quad \cdots \quad g_m(x)\,)\,,\ \mathcal{G} = \mathrm{span}\{G(x)\},$$

$$\mathcal{K}_i = \ker\{dh_i\}\ ,\ h(x) = (\, h_1(x) \quad \cdots \quad h_m(x)\,)^T .$$

We assume that $G(x_0)$ has full rank equal to m and $dh_i(x_0)$ is a nonzero row vector.

For a given feedback law (1.2), let

$$\alpha(x, w) = (\, \alpha_1(x, w) \quad \cdots \quad \alpha_m(x, w)\,)^T\ ,\ \beta(x, w) = (\, \beta_1(x, w) \cdots \quad \beta_m(x, w)\,)\,,$$

$$\delta(x, w) = (\, \delta_1(x, w) \quad \cdots \quad \delta_m(x, w)\,)$$

and

$$u = (\, u_1 \quad \dots \quad u_m\,)^T\ ,\ v = (\, v_1 \quad \dots \quad v_m\,)^T .$$

Moreover, denoting by $A(x)B(x)$ the matrix product between two given matrices $A(x)$ and $B(x)$, let

$$\widetilde{f}^e(x, w) = \begin{pmatrix} f(x) + G(x)\alpha(x, w) \\ \gamma(x, w) \end{pmatrix},$$

$$\widetilde{g}_j^e(x, w) = \begin{pmatrix} G(x)\beta_j(x, w) \\ \delta_j(x, w) \end{pmatrix}, \qquad j = 1, \dots, m,$$

$$\widetilde{G}^e(x, w) = (\, \widetilde{g}_1^e(x, w) \quad \cdots \quad \widetilde{g}_m^e(x, w)\,).$$

The notations x^e and x_0^e will be used to denote (x, w) and $(x_0, w_0) = (0, 0)$, respectively. In the case $n^w = 0$, we obtain *static* feedback laws and, more simply, we replace $\widetilde{f}^e(x, w)$ by $\widetilde{f}(x)$, $\widetilde{g}_j^e(x, w)$ by $\widetilde{g}_j(x)$ and $\widetilde{G}^e(x, w)$ by $\widetilde{G}(x)$.

Given a smooth vector field d and a matrix $A(x)$ (with suitable dimensions), we also denote by $L_A d(x)$ the matrix $\frac{\partial d}{\partial x}(x)A(x)$, where $\frac{\partial d}{\partial x}(x)$ is the Jacobian of $d(x)$ at x. Moreover, if $a_{ij}(x)$ is the (i, j)–th entry of $A(x)$, by $L_d A(x)$ we denote the matrix with $(\partial a_{ij}/\partial x)(x)d(x)$ as (i, j)–th entry.

Given a smooth distribution Δ and a smooth vector field d, we denote by $[d, \Delta]$ the smooth distribution spanned by the vector fields $\{[d, \tau] : \tau \in \Delta\}$. If $d_1, \ldots, d_s$ and $\tau_1, \ldots, \tau_r$ are smooth vector fields and $D(x)$ and $T(x)$ are the matrices $(\, d_1(x) \quad \cdots \quad d_s(x)\,)$ and $(\, \tau_1(x) \quad \cdots \quad \tau_r(x)\,)$, respectively, $[D, T](x)$ will denote the matrix

$$(\, [d_1, \tau_1](x) \quad \cdots \quad [d_s, \tau_1](x) \quad \cdots \quad [d_1, \tau_r](x) \quad \cdots \quad [d_s, \tau_r](x) \,) \, .$$

3.3 Noninteracting control with local stability via static feedback

Let us consider the class of regular feedback laws $u = \alpha(x) + \beta(x)v$. A necessary and sufficient condition for NS to be solvable (via regular feedback) is that (1.1) have vector relative degree at x_0 ([55], section 5.3). A characterization of this condition in geometric terms has been given in [62] and [64] and will be used in section 4.3 to approach the problem of noninteracting control with stability in the case of block–partitioned outputs. As a consequence of the above remarks, the following assumption arises naturally when investigating necessary and sufficient conditions to solve NLSS.

(A1) The system (1.1) has vector relative degree at x_0.

Note that, if assumption (A1) holds, any invertible feedback law $u = \alpha(x) + \beta(x)v$, which solves NS, is also regular. For this reason, in what follows, we will consider, more generally, the class of *invertible* static feedback laws. Under assumption (A1), a simple invertible feedback law, which solves NS, is given by

$$u = -A^{-1}(x)D(x) + A^{-1}(x)v, \tag{3.1}$$

where $D(x) = (\, L_f^{r_1-1}h_1(x) \quad \cdots \quad L_f^{r_m-1}h_m(x)\,)^T$ and $A(x)$ is the decoupling matrix of (1.1) ([55], section 5.3). Note that the system (1.1)–(3.1), after a suitable change of coordinates, appears as m chains of r_i integrators each plus an unobservable dynamics.

Now, let $\mathcal{V}_i^*$ and $\mathcal{R}_i^*$ be, respectively, the maximal locally controlled invariant distribution for (1.1) contained in $\bigcap_{j\neq i}\mathcal{K}_j$ and the maximal local controllability distribution for (1.1) contained in $\bigcap_{j\neq i}\mathcal{K}_j$, both supposed to exist. Moreover, let $\mathcal{R}_0$ be the smallest distribution which is locally invariant under $f, g_1, \dots, g_m$ and contains $\mathcal{G}$ (this distribution always exists). These distributions can be computed by means of the algorithms given in chapter 1 and are said to be *regularly computable* (at x_0) if the distributions given at each step by these algorithms have constant dimension in a neighbourhood of x_0. Moreover, let $\mathcal{R}^* = \bigcap_{i=1}^{m} \sum_{j\neq i}\mathcal{R}_j^*$. We will assume that

(A2) the distributions $\mathcal{V}_i^*$, $\mathcal{R}_i^*$, $i = 1, \dots, m$ and $\mathcal{R}_0$ are regularly computable at x_0. Moreover, the distributions $\sum_{j\in I}\mathcal{R}_j^*$, $\mathcal{R}^*$, $(\sum_{j\in I}\mathcal{R}_j^*)\cap\mathcal{G}$ and $\sum_{j\in I}(\mathcal{R}_j^*\cap\mathcal{G})$, $I\subset\{1,\dots,m\}$, have constant dimension in a neighbourhood of x_0.

Moreover,

(A3) $\dim\mathcal{R}_0(x) = n$ *for all x in a neighbourhood of x_0.*

This is the well-known *strong-accessibility* assumption ([83]) and, for linear systems, amounts to the pair (A, B) being controllable. Unlike linear systems, this condition does not imply the existence of a feedback law (even continuous) which stabilizes the system at the origin of the state space (see [80] for a very intuitive explanation). A system (1.1), which satisfies assumption (A3), is said to be strong accessible (at x_0). If $\dim\mathcal{R}_0(x) < n$, by Frobenius' theorem (see [55], theorem **1.4.1**, for an easy introduction) there exists a change of coordinates $z(x) = (\, z_1^T(x) \quad z_2^T(x)\,)^T$, $z(x_0) = 0$, defined in a neighbourhood of x_0 and such that (1.1) is expressed in these coordinates by

$$\begin{aligned} \dot z_1 &= f_1(z_1, z_2) + \sum_{j=1}^{m} g_{1j}(z_1, z_2)u_j, \\ \dot z_2 &= f_2(z_2), \\ y_i &= h_i(z_1, z_2), \qquad i = 1, \dots, m, \end{aligned} \tag{3.2}$$

with $\mathcal{R}_0 = \mathrm{span}\{\partial/\partial z_1\}$. Note that by construction the system

$$\begin{aligned} \dot z_1 &= f_1(z_1, 0) + \sum_{j=1}^{m} g_{1j}(z_1, 0)u_j, \\ y_i &= h_i(z_1, 0), \qquad i = 1, \dots, m, \end{aligned} \tag{3.3}$$

is strong-accessible at $z_1 = 0$. Clearly, if $\dot z_2 = f_2(z_2)$ is unstable in $z_2 = 0$, NLSS is not solvable. On the other hand, if $\dot z_2 = f_2(z_2)$ is locally asymptotically stable in $z_2 = 0$ and

$z_2(t, \bar{z}_2)$, the trajectory of $\dot{z}_2 = f_2(z_2)$ at time $t \geq 0$ with initial state $\bar{z}_2$, goes to zero "sufficiently fast", then NLSS is solvable in an "approximate" sense. Let us state this formally.

Denote by $\|\cdot\|_2$ the euclidean norm of $I\!R^n$. Since by assumption $\dot{z}_2 = f_2(z_2)$ is locally asymptotically stable in $z_2 = 0$, $\forall\epsilon > 0$ there exist $T_\epsilon > 0$ and $\delta_\epsilon > 0$ such that $\forall t \geq T_\epsilon$ and $\|\bar{z}_2\|_2 < \delta_\epsilon$ we have $\|z_2(t, \bar{z}_2)\|_2 < \epsilon$, i.e. the trajectories of $\dot{z}_2 = f_2(z_2)$, starting inside an open ball with ray δ_ϵ, enter the closed ball with ray ϵ within a time T_ϵ. From proposition 1.1, it follows that, if $u = \alpha(z_1) + \beta(z_1)v$ is a feedback law which solves NLSS for (3.3), then the system (3.2) with $u = \alpha(z_1)$ is locally asymptotically stable in $z = 0$. Let $\widetilde{\Sigma}$ be the system (3.2) with $u = \alpha(z_1) + \beta(z_1)v$ and fix $i \in \{1, \dots, m\}$. Moreover, let $v_a^i(t)$ and $v_b^i(t)$ be input vectors with all the components but the i-th one equal to zero for all $t \geq t_0$ and $z(t, t_0, \bar{z}, v(t))$ be the trajectory of $\widetilde{\Sigma}$ at time $t \geq t_0$, with initial state $\bar{z}$ and input $v(t)$. By continuity, it follows that for each $\epsilon' > 0$ there exist $T_{\epsilon',t_0} > 0$, $\delta_{\epsilon',t_0} > 0$ and $\eta_{\epsilon',t_0} < 0$ such that for $t \geq T_{\epsilon',t_0} \geq t_0$ we have $|y_j(z(t, t_0, \bar{z}, v_a^i(t))) - y_j(z(t, t_0, \bar{z}, v_b^i(t)))| < \epsilon'$ for $j \neq i$, as long as $\|\bar{z}\|_2 < \delta_{\epsilon',t_0}$, $\|v_a^i(t)\|_2 < \eta_{\epsilon',t_0}$ and $\|v_b^i(t)\|_2 < \eta_{\epsilon',t_0}$ for all $t \geq t_0$. Thus, if T_{ϵ',t_0} is sufficiently small, $\widetilde{\Sigma}$ is *approximately* noninteractive in the sense that $y_j(z(t, t_0, \bar{z}, v_a^i(t))$ and $y_j(z(t, t_0, \bar{z}, v_b^i(t))$ are arbitrarily close for $j \neq i$ and for sufficiently small inputs $v_a^i(t)$, $v_b^i(t)$ and initial states $\bar{z}$.

3.3.1 Invariant dynamics in noninteracting control via static feedback

In this section we will show that it is always possible to *uniquely* associate some nonlinear dynamics (to be defined) to a given system (1.1), which satisfies assumptions (A1), (A2) and (A3). These dynamics are crucial in finding necessary and sufficient conditions to solve NLSS. First, we give some preliminary results.

Proposition 3.1. Assume (A1), (A2) and (A3) hold. Let us consider an invertible feedback law $u = \alpha(x) + \beta(x)v$, *which solves NS, and suppose that the distributions* $\langle \tilde{f}, \tilde{g}, \dots, \tilde{g}_m | \text{span}\{\tilde{g}_i\}\rangle$, $i = 1, \dots, m$, *have constant dimension in a neighbourhood of* x_0. *We have*

$$\sum_{j\in I} \mathcal{R}_j^* = \langle \tilde{f}, \tilde{g}_1, \dots, \tilde{g}_m | \, span\{\tilde{g}_j : j \in I\}\rangle, \qquad I \subset \{1, \dots, m\},$$

in a neighbourhood of x_0. *Thus,* $\sum_{j\in I} \mathcal{R}_j^*$ *is a local controllability distribution for (1.1) and, in particular, locally controlled invariant and involutive.* □

Proof. Using the same arguments contained in [55], proposition 7.4.2, it can be shown that $\mathcal{R}_i^* = \langle \tilde{f}, \tilde{g}_1, \dots, \tilde{g}_m | \text{span}\{\tilde{g}_i\}\rangle$, $i = 1, \dots, m$. Since by assumption the distributions $\sum_{j\in I} \mathcal{R}_j^*$

have constant dimension in a neighbourhood of x_0 and $\mathcal{R}_i^* = \langle \tilde{f}, \tilde{g}_1, \dots, \tilde{g}_m | \text{span}\{\tilde{g}_i\} \rangle$, it follows that also the distributions $\sum_{j \in I} \langle \tilde{f}, \tilde{g}_1, \dots, \tilde{g}_m | \text{span}\{\tilde{g}_j\} \rangle$ have constant dimension in a neighbourhood of x_0 and, thus, being equal to the sum of locally invariant distributions, are locally invariant under $\tilde{f}, \tilde{g}_1, \dots, \tilde{g}_m$.Moreover, $\sum_{j \in I} \mathcal{R}_j^* \supset \text{span}\{\tilde{g}_j : j \in I\}$. Thus, $\sum_{j \in I} \mathcal{R}_j^* \supset \langle \tilde{f}, \tilde{g}_1, \dots, \tilde{g}_m | \text{span}\{\tilde{g}_j : j \in I\} \rangle$ by definition. Conversely, $\sum_{j \in I} \mathcal{R}_j^* \subset \langle \tilde{f}, \tilde{g}_1, \dots, \tilde{g}_m | \text{span}\{\tilde{g}_j : j \in I\} \rangle$, since by definition

$$\langle \tilde{f}, \tilde{g}_1, \dots, \tilde{g}_m | \text{span}\{\tilde{g}_j\} \rangle \subset \langle \tilde{f}, \tilde{g}_1, \dots, \tilde{g}_m | \text{span}\{\tilde{g}_j : j \in I\} \rangle, \qquad j \in I.$$

Thus, $\sum_{j \in I} \mathcal{R}_j^* = \langle \tilde{f}, \tilde{g}_1, \dots, \tilde{g}_m | \text{span}\{\tilde{g}_j : j \in I\} \rangle$ in a neighbourhood of x_0. From lemma 2.8.8 of [55], since $\sum_{j \in I} \mathcal{R}_j^*$ has constant dimension in a neighbourhood of x_0, it follows that $\sum_{j \in I} \mathcal{R}_j^*$ is also locally involutive and, thus, by definition, it is a local controllability distribution for (1.1). □

In other words, the distributions $\sum_{j \in I} \mathcal{R}_j^*$, $I \subset \{1, \dots, m\}$, are locally involutive and rendered locally invariant by *any* invertible feedback law which solves NS. In particular, the distribution $\mathcal{R}^*$ is locally involutive, since it is equal to the intersection of involutive distributions, and rendered locally invariant by *any* invertible feedback law which solves NS. Since locally involutive, the distributions $\mathcal{R}_i^*$, $\mathcal{R}^*$ and $\sum_{j \neq i} \mathcal{R}_j^*$, $i = 1, \dots, m$, locally admit a foliation of the state space. The following lemma gives a characterization of the invertible feedback laws which solve NS and incorporates a result given in [54].

Proposition 3.2. Assume that (A1), (A2) and (A3) hold. Let $u = \alpha^h(x) + \beta^h(x)v$, $h = 1, 2$, be two invertible feedback laws which solve NS. Then, $\alpha^2(x) - \alpha^1(x) = \beta^1(x)\alpha^0(x)$, where $\alpha^0(x)$ is a vector of smooth real-valued functions, constant along each leaf of $\sum_{j \neq i} \mathcal{R}_j^*$ and equal to 0 for $x \in \mathcal{L}_{x_0}^{\sum_{j \neq i} \mathcal{R}_j^*}$, and $\beta^2(x) = \beta^1(x)\beta^0(x)$, where $\beta^0(x)$ is a diagonal matrix, with smooth diagonal entries, constant along each leaf of $\sum_{j \neq i} \mathcal{R}_j^*$ and nonzero at x_0. In particular, $\alpha^1(x) = \alpha^2(x)$ for $x \in \mathcal{L}_{x_0}^{\mathcal{R}^*}$. □

Proof. From proposition 3.1, it follows that for $X_i \in \sum_{j \neq i} \mathcal{R}_j^*$

$$[f + G\alpha^h, X_i] \in \sum_{j \neq i} \mathcal{R}_j^*, \qquad h = 1, 2,$$

which implies

$$[G\beta^1(\beta^1)^{-1}(\alpha^2 - \alpha^1), X_i] \in \sum_{j \neq i} \mathcal{R}_j^*. \tag{3.4}$$

From proposition 3.1 it also follows that the i–th column of $G\beta^1$, denoted in what follows by $\tilde{g}_i^1$ is a vector field of $\mathcal{R}_i^*$. Moreover, this vector field spans a one–dimensional distribution which is linearly independent from $\sum_{j\neq i}\mathcal{R}_j^*$ at each x of a neighbourhood of x_0. Indeed, suppose that in each neighbourhood of x_0 there exists $\bar{x}$ such that the subspace spanned by $\tilde{g}_i^1(\bar{x})$ has nonzero intersection with $\sum_{j\neq i}\mathcal{R}_j^*(\bar{x})$. We will show that this gives a contradiction. For, since $\tilde{\Sigma}^1$, the system obtained by plugging $u = \alpha^1(x) + \beta^1(x)v$ into (1.1), has vector relative degree $r_1, \ldots, r_m$ at x_0, we have

$$\mathcal{D}_i^* = (\mathrm{span}\{dh_i, \ldots, dL_{\tilde{f}^1}^{r_i-1}h_i\})^\perp, \qquad i = 1, \ldots, m, \tag{3.5}$$

where $\mathcal{D}_i^*$ is the maximal locally controlled invariant distribution for (1.1) contained in $\mathcal{K}_i$ and $\tilde{f}^1(x) = f(x) + G(x)\alpha^1(x)$ (see lemma 6.3.13 of [55] for a proof). Moreover, the decoupling matrix of $\tilde{\Sigma}^1$ is invertible at each x in a neighbourhood of x_0 and can be written as

$$\begin{pmatrix} dL_{\tilde{f}^1}^{r_1-1}h_1(x) \\ \vdots \\ dL_{\tilde{f}^1}^{r_m-1}h_m(x) \end{pmatrix} (\tilde{g}_1^1(x) \quad \cdots \quad \tilde{g}_m^1(x)).$$

From (3.5) and the invertibility of $\beta^1(x)$, it follows that

$$\mathcal{D}_i^* \cap \mathrm{span}\{\tilde{g}_i^1\} = 0 \tag{3.6}$$

in a neighbourhood of x_0, since otherwise the decoupling matrix would be singular at x_0. Since from proposition 3.1 we conclude that $\sum_{j\neq i}\mathcal{R}_j^*$ is a local controllability distribution for (1.1) (in particular, a locally controlled invariant distribution for (1.1)), contained in $\sum_{j\neq i}\bigcap_{h\neq j}\mathcal{K}_h$, which, in turn, is contained in $\mathcal{K}_i$, it follows by definition that $\sum_{j\neq i}\mathcal{R}_j^* \subset \mathcal{D}_i^*$. It follows from (3.6) that $(\sum_{j\neq i}\mathcal{R}_j^*) \cap \mathrm{span}\{\tilde{g}_i^1\} = 0$ in a neighbourhood of x_0. This gives the claimed contradiction.

From (3.1) we obtain

$$[G\beta^1, X_i](\beta^1)^{-1}(\alpha^2 - \alpha^1) - (G\beta^1)(L_{X_i}((\beta^1)^{-1}(\alpha^2 - \alpha^1))) \in \sum_{j\neq i}\mathcal{R}_j^*,$$

which, since $[G\beta^1, X_i] \in \sum_{j\neq i}\mathcal{R}_j^*$ and $\tilde{g}_i^1(x)$ spans a distribution which is linearly independent from $\sum_{j\neq i}\mathcal{R}_j^*$ at each x of a neighbourhood of x_0, implies that the Lie derivative of the i–th component of $(\beta^1)^{-1}(\alpha^2 - \alpha^1)$ along any vector field $X_i \in \sum_{j\neq i}\mathcal{R}_j^*$ is zero. This, together with the fact that $(\beta^1)^{-1}(\alpha^2 - \alpha^1)(x_0) = 0$ (the equilibrium point is preserved after feedback), implies the first part of the proposition. The second part can be proved as follows.

Since, denoting by $\widetilde{g}_i^j(x)$ the i–th column of $G\beta^j(x)$, we have $\mathcal{R}_i^* \cap \mathcal{G} = \mathrm{span}\{\widetilde{g}_i^1\} = \mathrm{span}\{\widetilde{g}_i^2\}$, it follows that $\beta^2(x) = \beta^1(x)\beta^0(x)$, where $\beta^0(x)$ is a diagonal matrix with smooth entries, nonzero at x_0. Moreover,

$$\sum_{j\neq i} \mathcal{R}_j^* \ni [G\beta^2, X_i] = [G\beta^1, X_i]\beta^0 - (G\beta^1)L_{X_i}\beta^0.$$

Since $[G\beta^1, X_i] \in \sum_{j\neq i} \mathcal{R}_j^*$ and $(\sum_{j\neq i} \mathcal{R}_j^*) \cap \mathrm{span}\{\widetilde{g}_i^1\} = 0$ in a neighbourhood of x_0, it follows that the Lie derivative of the i–th diagonal entry of $\beta^0(x)$ along any vector field $X_i \in \sum_{j\neq i} \mathcal{R}_j^*$ is zero, i.e. the i–th diagonal entry of $\beta^0(x)$ is constant for $x \in \mathcal{L}_{x_0}^{\sum_{j\neq i} \mathcal{R}_j^*}$. □

Note that, when $\beta^1(x) = I_{m\times m}$ and $\alpha^1(x) = 0$, i.e. if (1.1) is noninteractive, proposition 3.2 states that, given any invertible feedback law $u = \alpha(x) + \beta(x)v$ which solves NS, the i–th component of $\alpha(x)$ is constant along each leaf of $\sum_{j\neq i} \mathcal{R}_j^*$ and equal to 0 for $x \in \mathcal{L}_{x_0}^{\sum_{j\neq i} \mathcal{R}_j^*}$ and $\beta(x)$ is a diagonal matrix, with smooth diagonal entries, constant along each leaf of $\sum_{j\neq i} \mathcal{R}_j^*$ and nonzero at x_0. Let $\alpha_i(x)$ be the i–th component of $\alpha(x)$ and $\beta_{ii}(x)$ be the i–th diagonal entry of $\beta(x)$. Using the identity $L_{[Y_i,Y_j]}\varphi(x) = L_{Y_i}L_{Y_j}\varphi(x) - L_{Y_j}L_{Y_i}\varphi(x)$, where Y_i and Y_j are smooth vector fields and φ is a smooth real–valued function, it is easy to show by induction that

$$\begin{aligned} L_{\widetilde{g}_i} \widetilde{D}\beta_{ii}(x) &= 0, \\ L_{\widetilde{g}_i} \widetilde{D}\alpha_i(x) &= 0, \end{aligned} \tag{3.7}$$

for all x in a neighbourhood of x_0, where $\widetilde{D}$ is an arbitrary composition of the Lie derivatives $L_{\widetilde{f}}$ and $L_{\widetilde{g}_j}$, $j = 1, \ldots, m$ (see the proof of proposition 4.2). This implies that the feedback law $u = \alpha(x) + \beta(x)v$ is weakly noninteractive. Thus, the class of weakly noninteractive feedback laws arises naturally in solving NLSS.

Now, let us apply to (1.1) any invertible feedback law which solves NS (for example, the standard feedback law (3.1)). We will denote the resulting system by $\widetilde{\Sigma}$. The following lemma allows us to express $\widetilde{\Sigma}$ in a *standard* noninteractive form.

Proposition 3.3. Assume that (A1), (A2) and (A3) hold. Then, there exists a change of coordinates $z(x) = (\, z_1^T(x) \quad \cdots \quad z_{m+1}^T(x)\,)^T$, $z(x_0) = 0$, *defined in a neighbourhood of* x_0

and such that $\widetilde{\Sigma}$ is expressed in these coordinates by

$$\begin{aligned}\dot{z}_i &= \widetilde{f}_i(z_i) + \widetilde{g}_{ii}(z_i)v_i, \qquad i = 1, \ldots, m,\\ \dot{z}_{m+1} &= \widetilde{f}_{m+1}(z) + \sum_{j=1}^{m} \widetilde{g}_{m+1,j}(z)v_j,\\ y_i &= h_i(z_i), \qquad i = 1, \ldots, m,\end{aligned} \tag{3.8}$$

with $\sum_{j\neq i} \mathcal{R}_j^* = \operatorname{span}\{\partial/\partial z_j : j \neq i\}$, $i = 1, \ldots, m$. *In particular,* $\mathcal{R}^* = \operatorname{span}\{\partial/\partial z_{m+1}\}$. □

Proof. The proof uses some standard arguments contained in [54]. From proposition 3.1 and assumption (A1), it follows that

$$\sum_{j\neq i} \mathcal{R}_j^* + \bigcap_{j\neq i}(\sum_{k\neq j} \mathcal{R}_k^*) = \sum_{j=1}^{m} \mathcal{R}_j^* = \mathbb{R}^n, \qquad i = 1, \ldots, m,$$

or, equivalently,

$$(\sum_{j\neq i} \mathcal{R}_j^*)^\perp \bigcap (\sum_{j\neq i}(\sum_{k\neq j} \mathcal{R}_k^*)^\perp) = 0, \qquad i = 1, \ldots, m.$$

In other words, the codistributions $(\sum_{j\neq i} \mathcal{R}_j^*)^\perp$, $i = 1, \ldots, m$, are linearly independent in a neigbourhood of x_0. Since $\sum_{j\neq i} \mathcal{R}_j^*$ is locally involutive, by Frobenius' theorem (see[52], pp. 25–30), there exist smooth functions $z_1(x), \ldots, z_m(x)$ (with z_i possibly vector–valued) such that $\operatorname{span}\{dz_i\} \cap \sum_{j\neq i} \operatorname{span}\{dz_j\} = 0$ and $(\sum_{j\neq i} \mathcal{R}_j^*)^\perp = \operatorname{span}\{dz_i\}$, $i = 1, \ldots, m$, in a neighbourhood of x_0. Moreover, we can choose $z_{m+1}(x)$ in such a way that the Jacobian of $z(x) = (\, z_1^T(x) \;\; \cdots \;\; z_{m+1}^T(x)\,)^T$ is nonsingular at x_0. By the implicit function theorem (see [7] or [55], appendix **A**]), it follows that $z = z(x)$ is a local diffeormorphism, i.e. a change of coordinates defined in a neighbourhood of x_0. Moreover,by proposition 3.1 $\mathcal{R}_i^*$ is locally invariant under $\widetilde{f}, \widetilde{g}_1, \ldots, \widetilde{g}_m$. Since $\sum_{j\neq i} \mathcal{R}_j^* = \operatorname{span}\{\partial/\partial z_j : j \neq i\}$ and $\sum_{j\neq i} \mathcal{R}_j^* \subset \mathcal{K}_i$, we have

$$\begin{aligned}\dot{z}_i &= \widetilde{f}_i(z_i) + \sum_{j=1}^{m} \widetilde{g}_{ij}(z_i)v_j, \qquad i = 1, \ldots, m,\\ \dot{z}_{m+1} &= \widetilde{f}_{m+1}(z) + \sum_{j=1}^{m} \widetilde{g}_{m+1,j}(z)v_j,\\ y_i &= h_i(z_i), \qquad i = 1, \ldots, m,\end{aligned}$$

Moreover, since $\widetilde{g}_i \in \mathcal{R}_i^*$ and $\mathcal{R}_i^* \subset \bigcap_{j\neq i} \sum_{k\neq j} \mathcal{R}_k^* = \operatorname{span}\{\partial/\partial z_i, \partial/\partial z_{m+1}\}$, it follows that $\widetilde{g}_{ij}(z_i) = 0$ for $j \neq i$, $i = 1, \ldots, m$. This completes the proof. □

In what follows, for simplicity of notation, in (3.8) we let $z = x$ and $v = u$. In analogy with section **2.2**, now we introduce some nonlinear dynamics, which play a fundamental role in the solution of NLSS. As already mentioned, since $\widetilde{\Sigma}$ is noninteractive, the distributions $\mathcal{R}_i^*$, $i = 1, \ldots, m$, and $\mathcal{R}^*$ are locally invariant under $\tilde{f}, \tilde{g}_1, \ldots, \tilde{g}_m$. Since the feedback law (3.1) preserves the equilibrium point and $\mathcal{L}_{x_0}^{\mathcal{R}^*}$ is well-defined, it makes sense to consider the restriction of $\tilde{f}$ to $\mathcal{L}_{x_0}^{\mathcal{R}^*}$, denoted by $\tilde{f}|\mathcal{L}_{x_0}^{\mathcal{R}^*}$. This vector field induces a dynamics evolving on $\mathcal{L}_{x_0}^{\mathcal{R}^*}$, which in coordinates (3.8) is given by

$$\dot{x}_{m+1} = \tilde{f}_{m+1}(0, \ldots, 0, x_{m+1}). \tag{3.9}$$

This dynamics is, in the linear case, exactly (**2**.2.3), since, as it will be shown in chapter **4**, $\mathcal{Q}^* = 0$ when $m = p = \nu$.

Note that $\mathcal{R}_i^* + \mathcal{R}^* = \text{span}\{\partial/\partial x_i, \partial/\partial x_{m+1}\}$ and that $\mathcal{R}_i^* + \mathcal{R}^*$ is locally involutive and invariant under $\tilde{f}, \tilde{g}_1, \ldots, \tilde{g}_m$. Indeed, $\mathcal{R}_i^* + \mathcal{R}^* \subset \bigcap_{j\neq i} \sum_{k\neq j} \mathcal{R}_k^* = \text{span}\{\partial/\partial x_i, \partial/\partial x_{m+1}\}$. Conversely, from proposition 3.1 it follows that $\sum_{i=1}^{m} \mathcal{R}_i^* = \mathcal{R}_0 = I\!R^n$. This, together with $\sum_{j\neq i} \mathcal{R}_j^* = \text{span}\{\partial/\partial x_j : j \neq i\}$, implies that $\text{span}\{\partial/\partial x_i\} \subset \mathcal{R}_i^* + \mathcal{R}^*$ and, consequently, $\text{span}\{\partial/\partial x_i, \partial/\partial x_{m+1}\} \subset \mathcal{R}_i^* + \mathcal{R}^*$.

For the same reasons above, it makes sense to consider the restriction of $\tilde{f}$ to $\mathcal{L}_{x_0}^{\mathcal{R}_i^*+\mathcal{R}^*}$. Since $\tilde{g}_i \in \mathcal{R}_i^*$, we can also consider the restriction of $\tilde{g}_i$ to $\mathcal{L}_{x_0}^{\mathcal{R}_i^*+\mathcal{R}^*}$. The vector fields $\tilde{f}|\mathcal{L}_{x_0}^{\mathcal{R}_i^*+\mathcal{R}^*}$ and $\tilde{g}_i|\mathcal{L}_{x_0}^{\mathcal{R}_i^*+\mathcal{R}^*}$ induce a control system evolving on $\mathcal{L}_{x_0}^{\mathcal{R}_i^*+\mathcal{R}^*}$, which in coordinates (3.8) is given by

$$\begin{aligned}
\dot{x}_i &= \tilde{f}_i(x_i) + \tilde{g}_{ii}(x_i)v_i, \\
\dot{x}_{m+1} &= \tilde{f}_{m+1}(0, \ldots, 0, x_i, 0, \ldots, 0, x_{m+1}) + \tilde{g}_{m+1,i}(0, \ldots, 0, x_i, 0, \ldots, 0, x_{m+1})v_i.
\end{aligned}$$

The dynamics

$$\dot{x}_i = \tilde{f}(x_i) + \tilde{g}_{ii}(x_i)v_i \tag{3.10}$$

is, in the linear case, exactly (**2**.2.4).

The following proposition states that the dynamics (3.9) and (3.10) are *uniquely* (in some sense to be specified) associated to a given system (1.1), which satisfies assumptions (A1), (A2) and (A3). This renders the definition of (3.9) and (3.10) *independent* (in some sense) of the particular feedback law chosen to obtain (3.8). Given the system (3.8), let us apply to it any invertible feedback law $u = \alpha'(x) + \beta'(x)v$ which solves NS. From proposition 3.2 it follows that $\beta'(x)$ is a diagonal matrix. Let

$$\alpha'(x) = \begin{pmatrix} \alpha_1'(x) \\ \vdots \\ \alpha_m'(x) \end{pmatrix}, \quad \beta'(x) = \begin{pmatrix} \beta_{11}'(x) & \cdots & 0 \\ \vdots & \cdots & \vdots \\ 0 & \cdots & \beta_{mm}'(x) \end{pmatrix}.$$

and denote by $\widetilde{\Sigma}'$ the system obtained from (3.8) and by (3.9)′ and (3.10)′ the systems defined on $\widetilde{\Sigma}'$ and corresponding to (3.9) and (3.10), respectively.

Proposition 3.4. The dynamics (3.9) and (3.9)′ are equal up to changes of coordinates, defined in a neighbourhood of x_0. The systems (3.10) and (10)′ are equal up to changes of coordinates and invertible static feedback transformations, defined in a neighbourhood of x_0. □

Proof. Choose $\alpha^1(x) = 0$, $\beta^1(x) = I_{m\times m}$, $\alpha^2(x) = \alpha'(x)$ and $\beta^2(x) = \beta'(x)$. From proposition 3.2 it follows that $\alpha'(x) = 0$ for $x \in \mathcal{L}_{x_0}^{\mathcal{R}^*}$. Moreover, the controllability distributions $\mathcal{R}_1^*, \ldots, \mathcal{R}_\nu^*$ (and, thus, $\mathcal{R}^*$) are invariant under regular feedback laws ([55], lemma 6.4.3). This implies that (3.9) and (3.9)′ are equal up to local diffeomorphisms. Now, we prove the second part of the proposition. Let $\widetilde{\Sigma}$ be given by

$$\dot{x} = \widetilde{f}(x) + \sum_{i=1}^{m} \widetilde{g}_i(x)u_i,$$

and $\widetilde{\Sigma}'$ by

$$\dot{x} = \widetilde{f}'(x) + \sum_{i=1}^{m} \widetilde{g}_i'(x)v_i.$$

From (3.7) it follows that the functions $\alpha_i'(x)$ and $\beta_{ii}'(x)$ depend only on x_i and $\mathcal{R}_i^*$ is invariant under regular feedback laws, which are regular at x_0, the system (3.10)′ can be written as

$$\dot{x}_i = \widetilde{f}_i'(x_i) + \widetilde{g}_{ii}'(x_i)v_i = \widetilde{f}_i(x_i) + \widetilde{g}_{ii}(x_i)\alpha_i'(x_i) + \widetilde{g}_{ii}(x_i)\beta_{ii}'(x_i)v_i. \tag{3.11}$$

Since $\beta'(x_0) \neq 0$, by direct inspection of (3.11), it follows that (3.10) and (3.10)′ are equal up to invertible static feedback transformations (and change of coordinates), defined in a neighbourhood of x_0. □

From proposition 3.4 it follows that the local stability properties of (3.9) and (3.9)′ and the local stabilizability properties of (3.10) and (3.10)′, respectively, are the same.

3.3.2 Necessary and sufficient conditions

We are ready now to derive necessary and sufficient conditions to solve NLSS.

Let us begin with necessary conditions. Under assumptions (A1), (A2) and (A3), we will show that, if NLSS is solvable, then the dynamics (3.9) is locally asymptotically stable at the origin and each system (3.10) is locally asymptotically stabilizable at the origin via static feedback. To this end, we need first some auxiliary results.

From proposition 3.2 with $\alpha^1(x) = 0$ and $\beta^1(x) = I_{m\times m}$, for any invertible feedback law $u = \alpha(x) + \beta(x)v$, which solves NLSS for (3.8), we have $\alpha(x) = 0$ for $x \in \mathcal{L}_{x_0}^{\mathcal{R}^*}$. Since $\dot{x} = (\widetilde{f} + \widetilde{G}\alpha)(x)$ is locally asymptotically stable in x_0, then also (3.9) must be locally asymptotically stable at the origin. From (3.7) it follows that $\alpha_i(x)$ depends only on x_i. Since $\dot{x} = (\widetilde{f} + \widetilde{G}\alpha)(x)$ is locally asymptotically stable in x_0, also

$$\begin{aligned}
\dot{x}_i &= \widetilde{f}_i(x_i) + \widetilde{g}_{ii}(x_i)\alpha_i(x_i),\\
\dot{x}_{m+1} &= \widetilde{f}_{m+1}(0,\dots,0,x_i,0,\dots,x_{m+1}) + \widetilde{g}_{m+1,i}(0,\dots,0,x_i,0,\dots,x_{m+1})\alpha_i(x_i)
\end{aligned}$$

is locally asymptotically stable at the origin. This implies that (3.10) is locally asymptotically stabilizable at the origin via static feedback (pick $u_i = \alpha_i(x_i)$).

Conversely, the local asymptotic stability of (3.9) at the origin together with the local asymptotic stabilizability of (3.9) at the origin via static feedback are sufficient conditions to solve NLSS.Indeed, if $u_i = \alpha_i(x_i)$ locally asymptotically stabilizes (3.10) at the origin, the overall system

$$\begin{aligned}
\dot{x}_i &= \widetilde{f}_i(x_i) + \widetilde{g}_{ii}(x_i)\alpha_i(x_i) + \widetilde{g}_{ii}(x_i)v_i, \qquad i = 1,\dots,m,\\
\dot{x}_{m+1} &= \widetilde{f}_{m+1}(x) + \sum_{j=1}^{m}\widetilde{g}_{m+1,j}(x)\alpha_j(x_j) + \sum_{j=1}^{m}\widetilde{g}_{m+1,j}(x)v_j,\\
y_i &= h_i(x_i), \qquad i = 1,\dots,m,
\end{aligned} \tag{3.12}$$

is locally asymptotically stable at the origin. This is an easy consequence of proposition 1.1, since $\dot{x}_i = \widetilde{f}_i(x_i) + \widetilde{g}_{ii}(x_i)\alpha_i(x_i)$ and (3.9) are locally asymptotically stable at the origin and, moreover, $\alpha(x_0) = 0$. Clearly, (3.12) is also noninteractive, since the feedback law $u_i = \alpha_i(x_i) + v_i$, $i = 1,\dots,m$, preserves the triangular structure of (3.8).

We sum up the above results in the following theorem.

Theorem 3.5. Assume that (A1), (A2) and (A3) hold and consider the class of invertible feedback laws. If

a) the dynamics (3.9) is locally asymptotically stable at the origin,

b) each system (3.10) is locally asymptotically stabilizable at the origin via static feedback,

then NLSS is solvable. Conversely, assuming in addition that the distributions $\langle \widetilde{f}, \widetilde{g}_1, \dots, \widetilde{g}_m | \operatorname{span}\{\widetilde{g}_i\}\rangle$, $i = 1,\dots,m$, *have constant dimension in a neighbourhood of* x_0, *if NLSS is solvable, then a) and b) are satisfied.* □

Remark 3.6. In the case of linear systems (2.1.1) with $m = p = \nu$, assumption (A2) is trivially satisfed. On the other hand, (A1) and (A3) become respectively:

(A1L) the decoupling matrix of (2.1.1) is invertible,

(A3L) the pair (A, B) is controllable. □

As already pointed out in section 2.2, also condition b) of theorem 3.5 is automatically satisfied. Since (A1) and $\sum_{i=1}^{m} \mathcal{R}_i^* \cap \mathcal{B} = \mathcal{B}$ are equivalent conditions to achieve noninteraction via invertible static feedback laws [26], we recover the results of section 1.2.2. □

Remark 3.7. Assumption (A2) requires that $\mathcal{V}_i^*$, $\mathcal{R}_i^*$, $i = 1, \ldots, m$ and $\mathcal{R}_0$ be regularly computable at x_0. The results of theorem 3.5 are still true if we simply assume that the above distributions exist and have constant dimension in a neighbourhood of x_0. □

3.4 Noninteracting control with local stability via dynamic feedback

In this section, we will go through the solution of NLSD in two steps. First, in section 3.4.2 we will solve the problem for the class of systems which satisfy assumption (A1), (A2) and (A3). By doing this, we are in a position to start our analysis directly from the *standard* noninteractive form (3.8) and to give a very simple explicit expression of a dynamic feedback law which solves NLSD. The example of a system, which satisfies (A1), (A2) and (A3) and for which NLSD *is* solvable, will be given in section 3.5. Finally, in section 3.4.4 we will relax assumption (A1) and show that, under mild assumptions, for the class of systems (1.1) for which ND is solvable, there exists a *canonical* dynamic extension of (1.1) (denoted by Σ_C^q), which satisfies assumption (A1). This dynamic extension is *canonical*, in the sense that any other system, obtained from (1.1) after a feedback transformation and satisfying assumption (A1), can be obtained from Σ_C^q through invertible feedback laws and changes of coordinates. Thus, in order to solve NLSD for (1.1), first we extend canonically (1.1) to obtain Σ_C^q (when possible), which satisfies assumption (A1), and finally apply to Σ_C^q the results of section 3.4.2.

As mentioned above, we will start our analysis under assumptions (A1), (A2) and (A3). By doing this, we are in a privileged position. First, in analogy with section 3.3, we can assume that (1.1) is already in the form (3.8) (with $z = x$ and $v = u$). Secondly, by proposition 1.5, any feedback law which solves ND for (3.8) must be *weakly noninteractive*. This shows that the class of invertible weakly noninteractive feedback laws arises naturally in solving NLSD under assumptions (A1), (A2) and (A3).

3.4.1 Invariant dynamics in noninteracting control via dynamic feedback

It would be very nice if, in analogy with section 3.3.1, we could define for a given system (1.1), satisfying assumptions (A1), (A2) and (A3), some nonlinear dynamics with the property of being *invariant* (in some sense to be specified) under dynamic feedback laws which solve ND. This is actually possible in the following way. Let $\mathcal{I}^*$ be the Lie ideal generated by the vector fields $\{[\tilde{g}_i, ad^k_{\tilde{f}}\tilde{g}_j] : k \geq 0, i \neq j, i,j = 1,\ldots,m\}$ in the Lie algebra generated by $\tilde{f}, \tilde{g}_1, \ldots, \tilde{g}_m$ and let

$$\Delta^*_{MIX} = \text{span}\{\tau : \tau \in \mathcal{I}\}. \tag{4.1}$$

Note that $\Delta^*_{MIX} \subset \sum_{j \neq i} \mathcal{R}^*_j$ for $i = 1,\ldots,m$ and, in particular, $\Delta^*_{MIX} \subset \mathcal{R}^*$. Indeed, by proposition 3.1, $\sum_{j \neq i} \mathcal{R}^*_j = \langle \tilde{f}, \tilde{g}_1, \ldots, \tilde{g}_m | \text{span}\{\tilde{g}_j : j \neq i\} \rangle$. Moreover, the vector fields $[\tilde{g}_i, ad^k_{\tilde{f}}\tilde{g}_j]$, $k \geq 0$, $i \neq j$, $i,j = 1,\ldots,m$, are contained in $\sum_{j \neq i} \mathcal{R}^*_j$, $i = 1,\ldots,m$, since $\sum_{j \neq i} \mathcal{R}^*_j$ contains $\tilde{g}_j$ for $j \neq i$ and is locally invariant under $\tilde{f}, \tilde{g}_1, \ldots, \tilde{g}_m$. By similar reasons, we conclude that any vector field of $\mathcal{I}^*$ (and, thus, of Δ_{MIX}) is also a vector field of $\sum_{j \neq i} \mathcal{R}^*_j$. Moreover, it can be seen that Δ^*_{MIX} is spanned by the Lie brackets of $\tilde{f}, \tilde{g}_1, \ldots, \tilde{g}_m$, in which appear both $\tilde{g}_i$ and $\tilde{g}_j$ with $j \neq i$. This distribution is *intrinsically* associated to the problem of noninteracting control, in the sense that it cannot be defined for single-input systems.

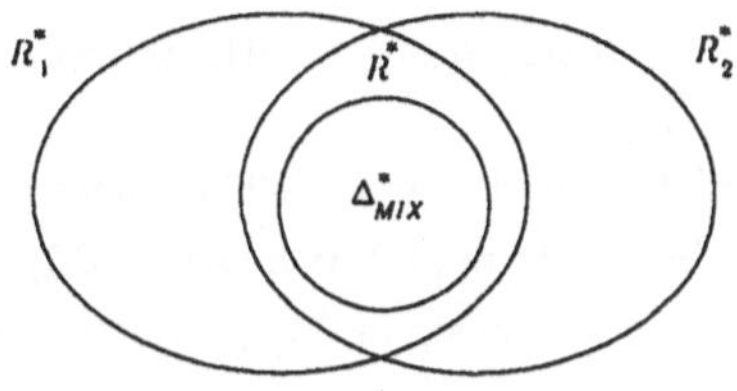

NONLINEAR SYSTEMS (m=p= 2)

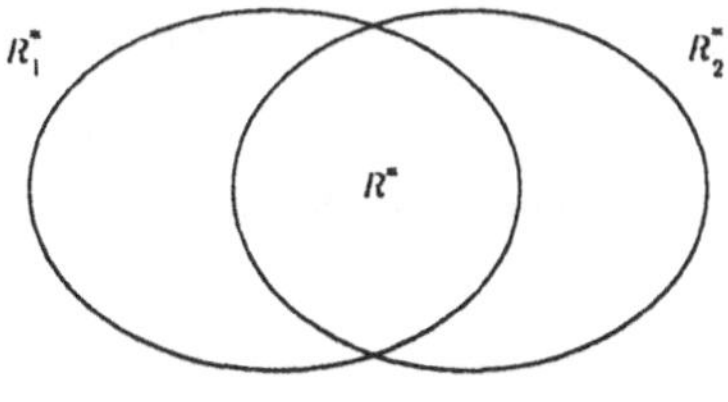

LINEAR SYSTEMS : $\Delta^*_{MIX} = 0$ (m=p= 2)

Figure 4.1

In the case of linear systems, since $\tilde{g}_i$ and $ad_{\tilde{f}}^k \tilde{g}_j$ are constant vector fields, we have $\Delta^*_{MIX} = 0$. Since, as we will see later, Δ^*_{MIX} is a nontrivial obstruction to solving NLSD, this fact is a natural consequence of theorem **2**.3.5 (see Figure 4.1).

To simplify our analysis, we will assume that

(A4) the distributions Δ^*_{MIX} and $\Delta^*_{MIX} + \mathcal{R}^*_i$, $i = 1, \ldots, m$ have constant dimension in a neighbourhood of x_0.

A consequence of this assumption and the definition of Δ^*_{MIX} is the following proposition.

Proposition 4.1. Under assumption (A1), (A2), (A3) and (A4), the distributions Δ^*_{MIX}, $\Delta^*_{MIX} + \mathcal{R}^*_i$ and $(\Delta^*_{MIX} + \mathcal{R}^*_i) \cap \mathcal{R}^*$, $i = 1, \ldots, m$, have constant dimension in a neighbourhood of x_0, are locally involutive and invariant under $\tilde{f}, \tilde{g}_1, \ldots, \tilde{g}_m$. □

Proof. Since Δ^*_{MIX} has constant dimension in a neighbourhood of x_0, there exist vector fields of the form

$$[\tilde{g}_{j_h}, [\ldots, [\tilde{g}_{j_2}, \tilde{g}_{j_1}] \ldots]], \, j_h, \ldots, j_1 \in \{0, \ldots, m\}, \, \tilde{g}_0(x) = \tilde{f}(x), \\ j_{k_1} \neq j_{k_2}, \, j_{k_1}, \, j_{k_2} \neq 0 \text{ for some } k_1, \, k_2 \in \{1, \ldots, h\}, \tag{4.2}$$

which are linearly independent at each x in a neighbourhood of x_0 and span Δ^*_{MIX}. The Lie bracket of any vector field of the form (4.2) (which is in Δ^*_{MIX} by definition) with either one of the vector fields $\tilde{f}, \tilde{g}_1, \ldots, \tilde{g}_m$ is again in Δ^*_{MIX} by definition. This proves that Δ^*_{MIX} is locally involutive and invariant under $\tilde{f}, \tilde{g}_1, \ldots, \tilde{g}_m$. Since $\mathcal{R}^*_i$, Δ^*_{MIX}, $\mathcal{R}^*_i + \Delta^*_{MIX}$, $\mathcal{R}^*$ and $\mathcal{R}^*_i + \mathcal{R}^*$ have costant dimension in a neighbourhood of x_0, our thesis follows from the fact that $\mathcal{R}^*_i$, $\mathcal{R}^*$ and Δ^*_{MIX} are locally involutive and invariant under $\tilde{f}, \tilde{g}_1, \ldots, \tilde{g}_m$. □

Using proposition 4.1 and reasoning as in the proof of proposition 3.3, it can be shown that there exists a change of coordinates

$$z^i(x) = (\, x_1^T \;\cdots\; x_{i-1}^T \;\; x_{i+1}^T \;\cdots\; x_m^T \;\; (z_1^i(x))^T \;\cdots\; (z_4^i(x))^T \,)^T, \tag{4.3}$$

$z^i(x_0) = 0$, defined in a neighbourhood of x_0 and such that the vector fields $\tilde{f}$ and $\tilde{g}_i$ are expressed in these coordinates by

$$\widetilde{f}(z^i) = \begin{pmatrix} \widetilde{f}_1(x_1) \\ \vdots \\ \widetilde{f}_{i-1}(x_{i-1}) \\ \widetilde{f}_{i+1}(x_{i-1}) \\ \vdots \\ \widetilde{f}_m(x_m) \\ \widetilde{f}_{z_1^i}(x_1,\ldots,x_{i-1},x_{i+1},\ldots,x_m,z_1^i) \\ \widetilde{f}_{z_2^i}(x_1,\ldots,x_{i-1},x_{i+1},\ldots,x_m,z_2^i) \\ \widetilde{f}_{z_3^i}(x_1,\ldots,x_{i-1},x_{i+1},\ldots,x_m,z_1^i,z_2^i,z_3^i) \\ \widetilde{f}_{z_4^i}(x_1,\ldots,x_{i-1},x_{i+1},\ldots,x_m,z_1^i,z_2^i,z_3^i,z_4^i) \end{pmatrix}, \tag{4.4}$$

$$\widetilde{g}_i(z) = \begin{pmatrix} 0 \\ \vdots \\ 0 \\ 0 \\ \vdots \\ 0 \\ \widetilde{g}_{iz_1^i}(x_1,\ldots,x_{i-1},x_{i+1},\ldots,x_m,z_1^i) \\ 0 \\ \widetilde{g}_{iz_3^i}(x_1,\ldots,x_{i-1},x_{i+1},\ldots,x_m,z_1^i,z_2^i,z_3^i) \\ \widetilde{g}_{iz_4^i}(x_1,\ldots,x_{i-1},x_{i+1},\ldots,x_m,z_1^i,z_2^i,z_3^i,z_4^i) \end{pmatrix}, \tag{4.5}$$

with

$$\begin{aligned} \mathcal{R}_i^* + \Delta_{MIX}^* &= \operatorname{span}\{\partial/\partial z_1^i, \partial/\partial z_3^i, \partial/\partial z_4^i\}, \\ \mathcal{R}^* &= \operatorname{span}\{\partial/\partial z_2^i, \partial/\partial z_3^i, \partial/\partial z_4^i\}, \\ \mathcal{R}^* \cap (\mathcal{R}_i^* + \Delta_{MIX}^*) &= \operatorname{span}\{\partial/\partial z_3^i, \partial/\partial z_4^i\}, \\ \Delta_{MIX}^* &= \operatorname{span}\{\partial/\partial z_4^i\}. \end{aligned} \tag{4.6}$$

Consider now the noninteractive system (3.8). We know from proposition 4.1 that the distributions $\mathcal{R}_i^* + \Delta_{MIX}^*$, $i = 1,\ldots,m$, and Δ_{MIX}^* have constant dimension in a neighbourhood of x_0, are locally involutive and invariant under $\widetilde{f}, \widetilde{g}_1, \ldots, \widetilde{g}_m$. Since $\widetilde{f}(x_0) = 0$ and $\mathcal{L}_{x_0}^{\Delta_{MIX}^*}$ is well-defined, it makes sense to consider the restriction $\widetilde{f}|\mathcal{L}_{x_0}^{\Delta_{MIX}^*}$. This vector field induces a dynamics evolving on $\mathcal{L}_{x_0}^{\Delta_{MIX}^*}$, which in z^i-coordinates is described by

$$\dot{z}_4^i = \widetilde{f}_{z_4^i}(0,\ldots,0,z_4^i). \tag{4.7}$$

This dynamics is trivial in the case of linear systems, since $\Delta_{MIX}^* = 0$.

It makes sense to consider also the restrictions $\widetilde{f}|\mathcal{L}_{x_0}^{\mathcal{R}_i^*+\Delta_{MIX}^*}$ and $\widetilde{g}_i|\mathcal{L}_{x_0}^{\mathcal{R}_i^*+\Delta_{MIX}^*}$ (note that $\widetilde{g}_i \in \mathcal{R}_i^*$). These vector fields define a control system evolving on $\mathcal{L}_{x_0}^{\mathcal{R}_i^*+\Delta_{MIX}^*}$, which in z^i-coordinates is given by

$$\begin{pmatrix} \dot{z}_1^i \\ \dot{z}_3^i \\ \dot{z}_4^i \end{pmatrix} = \begin{pmatrix} \widetilde{f}_{z_1^i}(0,\ldots,0,z_1^i) \\ \widetilde{f}_{z_3^i}(0,\ldots,0,z_1^i,0,z_3^i) \\ \widetilde{f}_{z_4^i}(0,\ldots,0,z_1^i,0,z_3^i,z_4^i) \end{pmatrix} + \begin{pmatrix} \widetilde{g}_{z_1^i}(0,\ldots,0,z_1^i) \\ \widetilde{g}_{iz_3^i}(0,\ldots,0,z_1^i,0,z_3^i) \\ \widetilde{g}_{iz_4^i}(0,\ldots,0,z_1^i,0,z_3^i,z_4^i) \end{pmatrix} u_i.$$

Let us consider the subsystem

$$\begin{pmatrix} \dot{z}_1^i \\ \dot{z}_3^i \end{pmatrix} = \begin{pmatrix} \widetilde{f}_{z_1^i}(0,\ldots,0,z_1^i) \\ \widetilde{f}_{z_3^i}(0,\ldots,0,z_1^i,0,z_3^i) \end{pmatrix} + \begin{pmatrix} \widetilde{g}_{iz_1^i}(0,\ldots,0,z_1^i) \\ \widetilde{g}_{iz_3^i}(0,\ldots,0,z_1^i,0,z_3^i) \end{pmatrix} u_i. \tag{4.8}$$

It is worth noting that, in the case of linear systems, since $\Delta_{MIX}^* = 0$ and $\mathcal{R}_i^*$ is a controllability subspace, (4.8) is a controllable system and, consequently, it is asymptotically stabilizable via static feedback.

One might ask if the dynamics (4.7) and (4.8) may depend on the feedback law chosen to obtain (3.8). The following proposition states that these dynamics are *uniquely* (in some sense) associated to a given system (1.1), which satisfies assumptions (A1), (A2), (A3) and (A4). This renders the definition of (4.7) and (4.8) *independent* (in some sense) of the particular feedback law chosen to obtain (3.8). Given the system (3.8), denoted by $\widetilde{\Sigma}$, let us apply to it any invertible feedback law $u = \alpha'(x) + \beta'(x)v$ which solves NS, with $\alpha'(x) = (\alpha_1'(x) \quad \cdots \quad \alpha_m'(x))^T$ and $\beta'(x) = (\beta_1'(x) \quad \cdots \quad \beta_m'(x))$. Denote by $\widetilde{\Sigma}'$ the system obtained from $\widetilde{\Sigma}$ and by (4.7)$'$ and (4.8)$'$ the dynamics defined on $\widetilde{\Sigma}'$ and corresponding to (4.7) and (4.8), respectively.

Proposition 4.2. The dynamics (4.7) and (4.7)$'$ are equal up to changes of coordinates, defined in a neighbourhood of x_0. The systems (4.8) and (4.8)$'$ are equal up to changes of coordinates and invertible static feedback transformations, defined in a neighbourhood of x_0. □

To prove proposition 4.2, we need a preliminary result. The proof of the first part of the following lemma can be found in [86], but, since it contains some ideas which will be often used in the sequel, we will repeat it here. Let $\beta_{ij}'(x)$ be the i-th component of $\beta_j'(x)$ and $(\Delta_{MIX}^*)'$ be the distribution defined as Δ_{MIX}^* but with $\widetilde{f}$ and $\widetilde{g}_j$, $j = 1,\ldots,m$, replaced by $\widetilde{f}' = \widetilde{f} + \widetilde{G}\alpha'$ and $\widetilde{g}_j' = \widetilde{G}\beta_j'$, respectively.

Lemma 4.3. Assume that (A1), (A2), (A3) and (A4) hold. Let $u = \alpha'(x) + \beta'(x)v$ be any invertible feedback law which solves NS for (3.8). Then, $\Delta_{MIX}^* = (\Delta_{MIX}^*)'$ in a *neighbourhood of* x_0. *Moreover,* $\alpha'_i(x) = 0$ *for* $x \in \mathcal{L}_{x_0}^{\mathcal{R}_j^* + \Delta^*_{MIX}}$ *and* $j \neq i$; *in particular,* $\alpha'(x) = 0$ *for* $x \in \mathcal{L}_{x_0}^{\Delta^*_{MIX}}$. □

Proof. From proposition 3.2 it follows that $\beta'_{ij}(x) = 0$ for all x in a neighbourhood of x_0 and $j \neq i$, $\beta'_{ii}(x_0) \neq 0$ and

$$\begin{aligned} L_{X_i}\beta'_{ii}(x) &= 0, \\ L_{X_i}\alpha'_i(x) &= 0, \end{aligned} \tag{4.9}$$

for $X_i \in \sum_{j\neq i} \mathcal{R}_j^*$ and for all x in a neighbourhood of x_0. Since $\alpha'(x_0) = 0$ and $\Delta^*_{MIX} \subset \sum_{j\neq i} \mathcal{R}_j^*$, (4.9) implies that $\alpha'_i(x) = 0$ for $x \in \mathcal{L}_{x_0}^{\mathcal{R}_j^* + \Delta^*_{MIX}}$ and $j \neq i$. In particular, $\alpha'(x) = 0$ for $x \in \mathcal{L}_{x_0}^{\Delta^*_{MIX}}$.

Now, we will prove that $\Delta^*_{MIX} = (\Delta^*_{MIX})'$ in a neighbourhood of x_0. We know that Δ^*_{MIX} is spanned by the Lie brackets of $\widetilde{f}, \widetilde{g}_1, \ldots, \widetilde{g}_m$, in which appear both $\widetilde{g}_i$ and $\widetilde{g}_j$ with $j \neq i$ and $i, j \in \{1, \ldots, m\}$. By using repeatedly the Jacobi identity, it can be easily shown that, equivalently, Δ^*_{MIX} is spanned by the vector fields of the form

$$[ad_{\widetilde{f}}^{s_q}\widetilde{g}_{i_q}, [\ldots, [ad_{\widetilde{f}}^{s_2}\widetilde{g}_{i_2}, ad_{\widetilde{f}}^{s_1}\widetilde{g}_{i_1}]\ldots]], \; s_i \geq 0, \; i_2 \neq i_1, \; i_1, \ldots, i_q \in \{1, \ldots, m\}. \tag{4.10}$$

Similarly, $(\Delta^*_{MIX})'$ is spanned by the vector fields of the form

$$[ad_{\widetilde{f}'}^{s_q}\widetilde{g}'_{i_q}, [\ldots, [ad_{\widetilde{f}'}^{s_2}\widetilde{g}'_{i_2}, ad_{\widetilde{f}'}^{s_1}\widetilde{g}'_{i_1}]\ldots]], \; s_i \geq 0, \; i_2 \neq i_1, \; i_1, \ldots, i_q \in \{1, \ldots, m\}. \tag{4.11}$$

From proposition 3.1 and using the identity $L_{[X_i, X_j]}\varphi(x) = L_{X_i}L_{X_j}\varphi(x) - L_{X_j}L_{X_i}\varphi(x)$, where X_i and X_j are smooth vector fields and φ is a smooth real-valued function, it is easy to show by induction that (4.9) is equivalent to

$$\begin{aligned} L_{X_i}\widetilde{D}\beta'_{ii}(x) &= 0, \\ L_{X_i}\widetilde{D}\alpha'_i(x) &= 0, \end{aligned} \tag{4.12}$$

for $X_i \in \sum_{j\neq i} \mathcal{R}_j^*$ and for all x in a neighbourhood of x_0, with $\widetilde{D}$ an arbitrary composition of the Lie derivatives $L_{\widetilde{f}}$ and $L_{\widetilde{g}_j}$, $j = 1, \ldots, m$. Indeed, from (4.9) and since $\mathcal{R}_i^*$ is invariant under $\widetilde{f}, \widetilde{g}_1, \ldots, \widetilde{g}_m$, it follows

$$\begin{aligned} L_{[\widetilde{f}, X_i]}\beta'_{ii}(x) &= L_{[\widetilde{g}_j, X_i]}\beta'_{ii}(x) = 0, \qquad j = 1, \ldots, m, \\ L_{[\widetilde{f}, X_i]}\alpha'_i(x) &= L_{[\widetilde{g}_j, X_i]}\alpha'_i(x) = 0, \qquad j = 1, \ldots, m, \end{aligned}$$

for all x in a neighbourhood of x_0. This, together with (4.9), implies $L_{X_i}L_{\widetilde{f}}\beta'_{ii}(x) = L_{X_i}L_{\widetilde{g}_j}\beta'_{ii}(x) = L_{X_i}L_{\widetilde{f}}\alpha'_i(x) = L_{X_i}L_{\widetilde{g}_j}\alpha'_i(x) = 0$ for all x in a neighbourhood of x_0. Now, suppose that for all x in a neighbourhood of x_0

$$\begin{aligned} L_{X_i}\widetilde{D}^k\beta'_{ii}(x) &= 0, \\ L_{X_i}\widetilde{D}^k\alpha'_i(x) &= 0, \end{aligned}$$

where $\widetilde{D}^k$ is an arbitrary composition of *at most* $k > 1$ Lie derivatives $L_{\widetilde{f}}$ and $L_{\widetilde{g}_j}$, $j = 1,\dots,m$. By induction hypothesis,

$$0 = L_{[\widetilde{f},X_i]}\widetilde{D}^k\beta'_{ii}(x) = L_{\widetilde{f}}L_{X_i}\widetilde{D}^k\beta'_{ii}(x) - L_{X_i}L_{\widetilde{f}}\widetilde{D}^k\beta'_{ii}(x) = \\ = -L_{X_i}L_{\widetilde{f}}\widetilde{D}^k\beta'_{ii}(x).$$

In the same way, we obtain $L_{X_i}L_{\widetilde{g}_j}\widetilde{D}^k\beta'_{ii}(x) = 0$ for all x in a neighbourhood of x_0 and the same equalities can be proved by replacing $\beta'_{ii}(x)$ with $\alpha'_i(x)$. This proves by induction that (4.12) implies (4.9). The converse is trivially true.

Now, for a given vector field of $\tau \in \mathcal{I}^*$, let r be the total number of $\widetilde{f}$'s and $\widetilde{g}_j$'s, $j = 1,\dots,m$, and s be the number of $\widetilde{f}$'s which appear in τ. Clearly, $r \geq s+2$. By induction on $q = r - s \geq 2$, we prove first that $(\Delta^*_{MIX})' \subset \Delta^*_{MIX}$ in a neighbourhood of x_0. By an easy induction argument on k, it follows that for $k \geq 0$ and $j_k + \dots + j_1 > 0$

$$ad^k_{\widetilde{f}'}\widetilde{g}'_i(x) = \beta'_{ii}(x)ad^k_{\widetilde{f}}\widetilde{g}_i(x) + \sum_{1\leq h\leq k}\ \sum_{j_h,\dots,j_1\in\{0,\dots,m\}} \varphi^{j_h\dots j_1 i}(x)[\widetilde{g}_{j_h},[\dots,[\widetilde{g}_{j_1},\widetilde{g}_i]\dots]](x), \tag{4.13}$$

where $\widetilde{g}_0(x) = \widetilde{f}(x)$ and $\varphi^{j_h\dots j_1 i}(x)$ are sums of terms of the following form

$$\begin{aligned} &D_h\cdots D_1\beta'_{ii}(x), && h\geq 1,\\ &D_t\cdots D_0(L_{ad^l_{\widetilde{f}'}\widetilde{g}'_i}\alpha'_i(x)), && 0\leq l+t<h, h\geq 1, l,t\geq 0, \end{aligned}$$

with $D_h,\dots,D_0 \in \{L_{\widetilde{f}'},\alpha'_1,\dots,\alpha'_m\}$. Moreover, in (4.13) only one of the two following facts can happen:

$$\begin{aligned} &-\ \varphi^{j_h\dots j_1 i}(x) \text{ contains no } \alpha'_j(x) \text{ for } j\neq i,\\ &-\ [\widetilde{g}_{j_h},[\dots,[\widetilde{g}_{j_1},\widetilde{g}_i]\dots]] \in \Delta^*_{MIX}. \end{aligned} \tag{4.14}$$

If $q = 2$, from (4.13) we have for $i_2 \neq i_1$, $l_{s_1} + \dots + l_1 > 0$, $j_{s_1} + \dots + j_1 > 0$ and $s_2 + s_1 \geq 0$

$$\begin{aligned} [ad^{s_2}_{\widetilde{f}'}\widetilde{g}'_{i_2}, ad^{s_1}_{\widetilde{f}'}\widetilde{g}'_{i_1}](x) &= \beta'_{i_2i_2}(x)\beta'_{i_1i_1}(x)[ad^{s_2}_{\widetilde{f}}\widetilde{g}_{i_2}, ad^{s_1}_{\widetilde{f}}\widetilde{g}_{i_1}](x) + \\ &+ \sum_{1\leq h\leq s_2}\ \sum_{1\leq k\leq s_1}\ \sum_{l_k,\dots,l_1\in\{0,\dots,m\}}\ \sum_{j_h,\dots,j_1\in\{0,\dots,m\}} \varphi^{j_h\dots j_1 i_2}(x)\varphi^{l_k\dots l_1 i_1}(x)\cdot\\ &\quad [[\widetilde{g}_{j_h},[\dots,[\widetilde{g}_{j_1},\widetilde{g}_{i_2}]\dots]],[\widetilde{g}_{l_k},[\dots,[\widetilde{g}_{l_1},\widetilde{g}_{i_1}]\dots]]](x) + \\ &- \sum_{1\leq h\leq s_2}\ \sum_{j_h,\dots,j_1\in\{0,\dots,m\}} (L_{ad^{s_1}_{\widetilde{f}'}\widetilde{g}'_{i_1}}\varphi^{j_h\dots j_1 i_2}(x)[\widetilde{g}_{j_h},[\dots,[\widetilde{g}_{j_1},\widetilde{g}_{i_2}]\dots]] + \\ &+ \sum_{1\leq k\leq s_1}\ \sum_{l_k,\dots,l_1\in\{0,\dots,m\}} (L_{ad^{s_2}_{\widetilde{f}'}\widetilde{g}'_{i_2}}\varphi^{l_k\dots l_1 i_1}(x)[\widetilde{g}_{l_k},[\dots,[\widetilde{g}_{l_1},\widetilde{g}_{i_1}]\dots]]. \end{aligned} \tag{4.15}$$

From (4.14) it follows that $[ad_{\widetilde{f}'}^{s_2}\widetilde{g}'_{i_2}, ad_{\widetilde{f}'}^{s_1}\widetilde{g}'_{i_1}] \in \Delta^*_{MIX}$ for $i_2 \neq i_1$. Suppose now that for some $q > 2$ and for $i_2 \neq i_1$

$$[ad_{\widetilde{f}'}^{s_q}\widetilde{g}'_{i_q}, [\ldots, [ad_{\widetilde{f}'}^{s_2}\widetilde{g}'_{i_2} ad_{\widetilde{f}'}^{s_1}\widetilde{g}'_{i_1}]\ldots]] \in \Delta^*_{MIX}.$$

If $\tau = [ad_{\widetilde{f}'}^{s_q}\widetilde{g}'_{i_q}, [\ldots, [ad_{\widetilde{f}'}^{s_2}\widetilde{g}'_{i_2}, ad_{\widetilde{f}'}^{s_1}\widetilde{g}'_{i_1}]\ldots]] \in \Delta^*_{MIX}$, from (4.12) and (4.13) and since Δ^*_{MIX} is locally invariant under $\widetilde{f}, \widetilde{g}_1, \ldots, \widetilde{g}_m$, we obtain for $s_{q+1} \geq 0$ and $j_{s_{q+1}} + \ldots + j_1 > 0$

$$\begin{aligned}[ad_{\widetilde{f}'}^{s_{q+1}}\widetilde{g}'_{i_{q+1}}, \tau](x) &= \beta'_{i_{q+1}i_{q+1}}(x)[ad_{\widetilde{f}}^{s_{q+1}}\widetilde{g}_{i_{q+1}}, \tau](x) + \\ &+ \sum_{1\leq h\leq s_{q+1}} \sum_{j_h,\ldots,j_1\in\{0,\ldots,m\}} (-L_\tau\varphi^{j_h\ldots j_1 i})(x)[\widetilde{g}_{j_h}, [\ldots, [\widetilde{g}_{j_1}, \widetilde{g}_{i_{q+1}}]\ldots]](x) + \\ &+ \sum_{1\leq h\leq s_{q+1}} \sum_{j_h,\ldots,j_1\in\{0,\ldots,m\}} \varphi^{j_h\ldots j_1 i}(x)[[\widetilde{g}_{j_h}, [\ldots, [\widetilde{g}_{j_1}, \widetilde{g}_{i_{q+1}}]\ldots]], \tau](x).\end{aligned}$$

From (4.14) and by induction, we conclude that $(\Delta^*_{MIX})' \subset \Delta^*_{MIX}$ in a neighbourhood of x_0.

Conversely, to prove that $(\Delta^*_{MIX})' \supset \Delta^*_{MIX}$ in a neighbourhood of x_0, we will use combined induction on $r \geq 2$ and $s \geq 0$. If $r = 2$ and $s = 0$, from (4.12) we obtain for $i_2 \neq i_1$

$$[\widetilde{g}'_{i_2}, \widetilde{g}'_{i_1}](x) = \beta'_{i_2 i_2}(x)\beta'_{i_1 i_1}(x)[\widetilde{g}_{i_2}, \widetilde{g}_{i_1}](x),$$

which, since $\beta(x_0) \neq 0$, implies $[\widetilde{g}_{i_2}, \widetilde{g}_{i_1}] \in (\Delta^*_{MIX})'$.

Now, suppose that for some $r > 2$ (with $s = 0$) and for $i_2 \neq i_1$

$$[\widetilde{g}_{i_r}, [\ldots, [\widetilde{g}_{i_2}, \widetilde{g}_{i_1}]\ldots]] \in (\Delta^*_{MIX})'. \tag{4.16}$$

As above, since $\beta(x_0) \neq 0$, it follows that

$$[\widetilde{g}_{i_{r+1}}, [\ldots, [\widetilde{g}_{i_2}, \widetilde{g}_{i_1}]\ldots]] \in (\Delta^*_{MIX})', \qquad i_2 \neq i_1.$$

By induction it follows that (4.16) holds for *any* $r \geq 2$ (with $s = 0$).

Assume that any vector field of $\mathcal{I}^*$ with *at most* s $\widetilde{f}$'s and *any* number $r \geq s + 2$ of $\widetilde{f}$'s and $\widetilde{g}_j$'s, $j = 1, \ldots, m$, is contained in $(\Delta^*_{MIX})'$. Let us consider any vector field $\tau \in \mathcal{I}^*$ with at most $s + 1$ $\widetilde{f}$'s and $r = s + 3$ $\widetilde{f}$'s and $\widetilde{g}_j$'s, $j = 1, \ldots, m$. Since in this case $q = r - s = s + 3 - (s + 1) = 2$, these Lie brackets are spanned by the vector fields of the form

$$[ad_{\widetilde{f}}^{s_2}\widetilde{g}_{i_2}, ad_{\widetilde{f}}^{s_1}\widetilde{g}_{i_1}](x), \qquad i_2 \neq i_1, \ 0 \leq s_2 + s_1 \leq s + 1.$$

Since the Lie brackets appearing on the right of (4.15) contain *at most* s $\widetilde{f}$'s (except for $[ad_{\widetilde{f}}^{s_2}\widetilde{g}_{i_2}, ad_{\widetilde{f}}^{s_1}\widetilde{g}_{i_1}]$) and $\beta(x_0) \neq 0$, by induction hypothesis it follows that

$$[ad_{\widetilde{f}}^{s_2}\widetilde{g}_{i_2}, ad_{\widetilde{f}}^{s_1}\widetilde{g}_{i_1}] \in (\Delta^*_{MIX})', \; i_2 \neq i_1 \,, \; 0 \leq s_2 + s_1 \leq s+1.$$

Similarly, by induction on $r > s+3$ with $s+1$ fixed, we can prove that for *any* $r \geq s+3$

$$\begin{gathered}[ad_{\widetilde{f}}^{s_q}\widetilde{g}_{i_q}, [\ldots, [ad_{\widetilde{f}}^{s_2}\widetilde{g}_{i_2}, ad_{\widetilde{f}}^{s_1}\widetilde{g}_{i_1}]\ldots]] \in (\Delta^*_{MIX})', \\ i_2 \neq i_1, s_q + \ldots s_1 \leq s+1, 2 \leq q \leq r-s-1.\end{gathered}$$

This proves by induction that $\Delta^*_{MIX} \subset (\Delta^*_{MIX})'$ in a neighbourhood of x_0. □

We are now in a position to prove proposition 4.2. From proposition 4.3, it follows that $\Delta^*_{MIX} = (\Delta^*_{MIX})'$ in a neighbourhood of x_0 and $\widetilde{f}(x) + \widetilde{g}(x)\alpha'(x) = \widetilde{f}(x)$ for $x \in \mathcal{L}_{x_0}^{\Delta^*_{MIX}}$. This implies the first part of proposition 4.2.

Now, consider $\widetilde{\Sigma}$ in z^i–coordinates. From the proof of propositions 3.2 and 3.3 and by (4.6), we conclude that $\alpha_i'(z^i)$ and $\beta_{ii}'(z^i)$ depend only on z_i^1. Moreover, $\Delta^*_{MIX} = (\Delta^*_{MIX})'$ in a neighbourhood of x_0 and $\mathcal{R}_i^*$ is invariant under regular feedback. Thus, (4.8)$'$ can be written as

$$\begin{aligned}\begin{pmatrix}\dot{z}_1^i \\ \dot{z}_3^i\end{pmatrix} &= \begin{pmatrix}\widetilde{f}'_{z_1^i}(0,\ldots,0,z_1^i) \\ \widetilde{f}'_{z_3^i}(0,\ldots,0,z_1^i,0,z_3^i)\end{pmatrix} + \begin{pmatrix}\widetilde{g}'_{iz_1^i}(0,\ldots,0,z_1^i) \\ \widetilde{g}'_{iz_3^i}(0,\ldots,0,z_1^i,0,z_3^i)\end{pmatrix} v_i = \\ &= \begin{pmatrix}\widetilde{f}_{z_1^i}(0,\ldots,0,z_1^i) \\ \widetilde{f}_{z_3^i}(0,\ldots,0,z_1^i,0,z_3^i)\end{pmatrix} + \begin{pmatrix}\widetilde{g}_{iz_1^i}(0,\ldots,0,0,z_1^i) \\ \widetilde{g}_{iz_3^i}(0,\ldots,0,z_1^i,0,z_3^i)\end{pmatrix} \alpha_i'(z_1^i) + \\ &+ \begin{pmatrix}\widetilde{g}_{iz_1^i}(0,\ldots,0,z_1^i) \\ \widetilde{g}_{iz_3^i}(0,\ldots,0,z_1^i,0,z_3^i)\end{pmatrix} \beta_{ii}'(z_1^i) v_i,\end{aligned} \tag{4.17}$$

with $\beta'(z^i(x_0)) \neq 0$. By direct inspection of (4.17), the second part of proposition 4.2 follows immediately.

Also in this case, from proposition 4.3 it follows that the local stability properties of (4.7) and (4.7)$'$ and, respectively, the local stabilizability properties of (4.8) and (4.8)$'$ are the same.

3.4.2 Necessary and sufficient conditions: an intermediate result

We are ready now to derive necessary and sufficient conditions to solve NLSD for the class of systems (1.1) which satisfy assumptions (A1), (A2), (A3) and (A4).

Let us begin with necessary conditions. Toward this end, a preliminary result is needed. We have seen in proposition 4.2 that the dynamics (4.7) is invariant under regular static feedback laws which solve NS. Let (1.2) be a feedback law which solves ND for (3.8). Moreover, let $(\Delta^*_{MIX})^e$ be defined as Δ^*_{MIX} but replacing $\widetilde{f}$ and $\widetilde{g}_j$, $j = 1, \ldots, m$, with $\widetilde{f}^e$ and $\widetilde{g}^e_j$ and $\mathcal{R}^e_i$ be the smallest distribution which is locally invariant under $\widetilde{f}^e, \widetilde{g}^e_1, \ldots, \widetilde{g}^e_m$ and contains span$\{\widetilde{g}^e_i\}$. Also, let π be the natural projection $\pi : \mathbb{R}^{n+n^w} \to \mathbb{R}^n$ and π_{*x^e} be the Jacobian of π at x^e. At this point, one might ask if (4.7) is also invariant, in the sense of *projection*, under feedback laws (1.2) which solve ND. This is answered in the next lemma.

Proposition 4.4. Assume that (A1), (A2), (A3) and (A4) hold. Let (1.2) be a feedback law which solves ND for (3.8) and assume that Δ^e_{MIX} have constant dimension in a neighbourhood of x^e_0. Then, for all x^e in a neighbourhood of x^e_0

$$\pi_{*x^e}\Delta^e_{MIX}(x^e) = \Delta^*_{MIX}(\pi(x^e)),$$

and $\alpha(x^e) = 0$ for $x^e \in \mathcal{L}^{\Delta^e_{MIX}}_{x^e_0}$. Moreover, if $\mathcal{R}^e_i$ is regularly computable at x^e_0, then for all x^e in a neighbourhood of x^e_0

$$\pi_{*x^e}(\mathcal{R}^e_i + \Delta^e_{MIX})(x^e) = (\mathcal{R}^*_i + \Delta^*_{MIX})(\pi(x^e))$$

and $\alpha_i(x^e) = 0$ for $x^e \in \mathcal{L}^{\mathcal{R}^e_j + \Delta^e_{MIX}}_{x^e_0}$ and $j \neq i$. □

Proof. The proof of the first part of the lemma can be found in [86]. We will sketch it here for completeness. From proposition 1.3 and proposition 1.5 it follows that in a neighbourhood of x^e_0 and for $X^e_i \in \sum_{j\neq i} \mathcal{R}^e_j$

$$\begin{aligned} L_{X^e_i}\widetilde{D}^e\beta_{ii}(x^e) &= 0, \\ L_{X^e_i}\widetilde{D}^e\alpha_i(x^e) &= 0, \\ \beta_{ij}(x^e) &= 0, \qquad j \neq i, \end{aligned} \tag{4.18}$$

where $\widetilde{D}^e$ is an arbitrary composition of the Lie derivatives $L_{\widetilde{f}^e}$ and $L_{\widetilde{g}^e_j}$, $j = 1, \ldots, m$. Moreover, from proposition 1.3 it follows that only one of the following facts can happen:

- $\beta_{ii}(x^e) \neq 0$ for all x^e in a neighbourhood of x^e_0;

- $\beta_{ii}(x^e) = 0$ for all x^e in a neighbourhood of x^e_0; moreover, there exists $\varrho_i \geq 1$ such that $L_{\widetilde{g}^e_i}L^k_{\widetilde{f}^e}\alpha_i(x^e) = 0$ for all x^e in a neighbourhood of x^e_0 and $k < \varrho_i - 1$ and

$L_{\tilde{g}_i^e} L_{\tilde{f}^e}^{\varrho_i - 1} \alpha_i(x^e) \neq 0$ for all x^e in a neighbourhood of x_0^e (in the case $\beta_{ii}(x^e) \neq 0$ for all x^e in a neighbourhood of x_0^e, we set $\varrho_i = 0$).

By an easy induction argument on k, it follows that for $k \geq \varrho_i$ and $j_{k-\varrho_i} + \ldots + j_1 > 0$

$$\pi_{*x^e}(ad_{\tilde{f}^e}^k \tilde{g}_i^e)(x^e) = \psi^i(x^e) ad_{\tilde{f}}^{k-\varrho_i} \tilde{g}_i(\pi(x^e)) + \\ + \sum_{1 \leq h \leq k-\varrho_i} \sum_{j_h, \ldots, j_1 \in \{0, \ldots, m\}} \varphi^{j_h \ldots j_1 i}(x^e) [\tilde{g}_{j_h}, [\ldots, [\tilde{g}_{j_1}, \tilde{g}_i] \ldots]](\pi(x^e)), \tag{4.19}$$

where

$$\psi^i(x^e) = \begin{cases} \beta_{ii}(x^e), & \text{if } \varrho_i = 0 \,, \\ -L_{ad_{\tilde{f}^e}^{\varrho_i - 1} \tilde{g}_i^e} \alpha_i(x^e), & \text{if } \varrho_i > 0 \,, \end{cases}$$

and $\varphi^{j_h \ldots j_1 i}(x^e)$ are sums of factors of the following form

$$D_h^e \cdots D_1^e \beta_{ii}(x^e), \qquad 1 \leq h \leq k - \varrho_i,$$
$$D_t^e \cdots D_0^e (L_{ad_{\tilde{f}^e}^l \tilde{g}_i^e} \alpha_i(x^e)), \qquad l + t < h \leq k - \varrho_i, h \geq 1, 0 \leq l, t \geq 0.$$

with $D_h^e \in \{L_{\tilde{f}^e}, \alpha_1, \ldots, \alpha_m\}$. From (4.18) and (4.19), proceeding as in the proof of proposition 4.2, we can show that $\pi_{*x^e} \Delta_{MIX}^e(x^e) = \Delta_{MIX}^*(\pi(x^e))$ for all x^e in a neighbourhood of x_0^e. Since $\Delta_{MIX}^e \subset \sum_{j \neq i} \mathcal{R}_j^e$ for $i = 1, \ldots, m$ and $\alpha(x_0^e) = 0$, it follows from (4.18) that $\alpha(x^e) = 0$ for $x^e \in \mathcal{L}_{x_0^e}^{\Delta_{MIX}^e}$.

To prove the second part of the lemma, it suffices to note that $\mathcal{R}_i^e + \Delta_{MIX}^e$ is spanned by vector fields of the form either

$$[\tilde{g}_{j_h}^e, [\ldots, [\tilde{g}_{j_1}^e, \tilde{g}_i^e] \ldots]], \\ j_h, \ldots, j_1 \in \{0, \ldots, m\}, \tilde{g}_0^e(x^e) = \tilde{f}^e(x^e),$$

or

$$[\tilde{g}_{j_h}^e, [\ldots, [\tilde{g}_{j_2}, \tilde{g}_{j_1}^e] \ldots]], \\ j_h, \ldots, j_1 \in \{0, \ldots, m\}, \tilde{g}_0^e(x^e) = \tilde{f}^e(x^e)\,, \\ j_{k_1} \neq j_{k_2} \text{ and } j_{k_1}, j_{k_2} \neq 0 \text{ for some } k_1\,,\ k_2 \in \{1, \ldots, h\},$$

and to proceed as above. □

Remark 4.5. Reasoning as in lemma 4.4, it is also possible to show that, in particular, $\pi_{*x^e} \mathcal{R}_i^e(x^e) = \mathcal{R}_i^*(\pi(x^e))$. Now, let us define *dynamic local controllability distribution* for (1.1) a locally involutive distribution $\Delta \subset I\!R^n$ for which there exist n^w and a feedback law (1.2), defined in a neighbourhood of x_0^e and such that

$$\pi_{*x^e} \langle \tilde{f}^e, \tilde{g}_1^e, \ldots, \tilde{g}_m^e | \text{span}\{\tilde{g}_i^e : i \in I\} \rangle (x^e) = \Delta(\pi(x^e)), \qquad I \subset \{1, \ldots, m\}.$$

Since $\pi_{*x^e}\mathcal{R}_i^e(x^e) = \mathcal{R}_i^*(\pi(x^e))$ by proposition 4.4, it follows that $\mathcal{R}_i^*$ is a dynamic local controllability distribution for (1.1). □

Suppose now that NLSD be solvable. Let π be as in proposition 4.4. It follows from this proposition that $\pi_{*x^e}\widetilde{f}^e(x^e) = \widetilde{f}(\pi(x^e))$ for $x^e \in \mathcal{L}_{x_0^e}^{\Delta_{MIX}^e}$. This exactly means that the trajectory of $\widetilde{\Sigma}^e$, starting from $\widehat{x}^e \in \mathcal{L}_{x_0^e}^{\Delta_{MIX}^e}$, and the trajectory of $\widetilde{\Sigma}$, starting from $\pi(\widehat{x}^e)$, have the same first n components. Thus, the dynamics (4.7) must be locally asymptotically stable at the origin, since $\dot{x}^e = \widetilde{f}^e(x^e)$ is (note that the trajectory of $\dot{x}^e = \widetilde{f}^e(x^e)$, starting from a given point of $\mathcal{L}_{x_0^e}^{\Delta_{MIX}^e}$, remains in $\mathcal{L}_{x_0^e}^{\Delta_{MIX}^e}$ by invariance of Δ_{MIX}^e).

Moreover, each system (4.8) must be locally asymptotically stabilizable at the origin via dynamic feedback, if NLSD is solvable. Indeed, assume that Δ_{MIX}^e, $\mathcal{R}_i^e$ and $\mathcal{R}_i^e + \Delta_{MIX}^e$, $i = 1,\ldots,m$, have constant dimension in a neighbourhood of x_0^e. Reasoning as in proposition 4.1, it can be shown that $\mathcal{R}_i^e + \Delta_{MIX}^e$ is locally involutive and invariant under $\widetilde{f}^e, \widetilde{g}_1^e, \ldots, \widetilde{g}_m^e$. Since $\alpha(x_0^e) = 0$, $\delta(x_0^e) = 0$ and $\mathcal{L}_{x_0^e}^{\mathcal{R}_i^e+\Delta_{MIX}^e}$ is well-defined, it makes sense to consider $\widetilde{f}^e|\mathcal{L}_{x_0^e}^{\mathcal{R}_i^e+\Delta_{MIX}^e}$. Moreover, from proposition 4.4, $\alpha_j(x^e) = 0$ for $x^e \in \mathcal{L}_{x_0^e}^{\Delta_{MIX}^e+\mathcal{R}_i^e}$ and $j \neq i$ and $\pi_{*x^e}(\mathcal{R}_i^e + \Delta_{MIX}^e)(x^e) = (\mathcal{R}_i^* + \Delta_{MIX}^*)(\pi(x^e))$ for all x^e in a neighbourhood of x_0^e. This implies that any vector field of $\mathcal{R}_i^e + \Delta_{MIX}^e$ projects to a well-defined vector field of $\mathcal{R}_i^* + \Delta_{MIX}^*$. In particular, this is true for $\widetilde{f}^e|\mathcal{L}_{x_0^e}^{\Delta_{MIX}^e+\mathcal{R}_i^e}$. If we consider (3.8) in z^i-coordinates and using the notation x^e also for the coordinates (z^i, w), since $\mathcal{R}_i^* + \Delta_{MIX}^* = \text{span}\{\partial/\partial z_1^i, \partial/\partial z_3^i, \partial/\partial z_4^i\}$, necessarily the component of the vector field $\pi_*(\widetilde{f}^e|\mathcal{L}_{x_0^e}^{\Delta_{MIX}^e+\mathcal{R}_i^e})$ along span $\{\partial/\partial x_j, \partial/\partial z_2^i : j \notin \{i, m+1\}\}$ is identically zero in a neighbourhood of $z^i(x_0)$. In other words, we obtain

$$\widetilde{f}^e|\mathcal{L}_{x_0^e}^{\mathcal{R}_i^e+\Delta_{MIX}^e} = \begin{pmatrix} \widetilde{f}_{z_1^i}(0,\ldots,0,z_1^i) \\ \widetilde{f}_{z_3^i}(0,\ldots,0,z_1^i,0,z_3^i) \\ \widetilde{f}_{z_4^i}(0,\ldots,0,z_1^i,0,z_3^i,z_4^i) \\ \varphi^i(z_1^i,z_3^i,z_4^i,w^*) \end{pmatrix} + \begin{pmatrix} \widetilde{g}_{iz_1^i}(0,\ldots,0,z_1^i) \\ \widetilde{g}_{iz_3^i}(0,\ldots,0,z_1^i,0,z_3^i) \\ \widetilde{g}_{z_4^i}(0,\ldots,0,z_1^i,0,z_3^i,z_4^i) \\ \eta^i(z_1^i,z_3^i,z_4^i,w^*) \end{pmatrix} \alpha_i(0,\ldots,0,z_1^i,0,z_3^i,z_4^i,w^*). \tag{4.20}$$

where $(z_1^i, z_3^i, z_4^i, w^*)$ is a set of local coordinates for $\mathcal{L}_{x_0^e}^{\mathcal{R}_i^e+\Delta_{MIX}^e}$ and $\varphi^i(z_1^i, z_3^i, z_4^i, w^*)$ and $\eta^i(z_1^i, z_3^i, z_4^i, w^*)$ are smooth (possibly vector-valued) functions of their arguments. Since $\dot{x}^e = \widetilde{f}^e(x^e)$ is locally asymptotically stable at x_0^e (note that the trajectory of $\dot{x}^e = \widetilde{f}^e(x^e)$, which starts from a given point of $\mathcal{L}_{x_0^e}^{\Delta_{MIX}^e+\mathcal{R}_i^e}$, remains in $\mathcal{L}_{x_0^e}^{\mathcal{R}_i^e+\Delta_{MIX}^e}$), by direct inspection of (4.20) it follows that also each system (4.8) must be locally asymptotically stable at the origin.

Conversely, if (4.7) is locally asymptotically stable at the origin and each system (4.8) is locally asymptotically stabilizable at the origin via dynamic feedback, then NLSD is solvable. We cannot prove this directly, so we will go through several steps. A first simplification of our problem is to assume $\Delta^*_{MIX} = 0$ in a neighbourhood of x_0 and we will see in a moment that this gives no loss of generality. Indeed, after a change of coordinates $z(x) = (z_1^T(x) \quad \cdots, z_m^T(x) \quad z_{m+1,1}^T(x) \quad z_{m+1,2}^T(x))^T$, $z(x_0) = 0$, defined in a neighbourhood of x_0, (3.8) becomes

$$\begin{aligned}
\dot{z}_i &= \widetilde{f}_i(z_i) + \widetilde{g}_{ii}(z_i)u_i, \qquad i = 1, \dots, m, \\
\dot{z}_{m+1,1} &= \widetilde{f}_{m+1,1}(z_1, \dots, z_m, z_{m+1,1}) + \sum_{j=1}^{m} \widetilde{g}_{m+1,1j}(z_1, \dots, z_m, z_{m+1,1})u_j, \\
\dot{z}_{m+1,2} &= \widetilde{f}_{m+1,2}(z) + \sum_{j=1}^{m} \widetilde{g}_{m+1,2j}(z)u_j, \\
y_i &= h_i(z_i), \qquad i = 1, \dots, m,
\end{aligned}$$

with $\sum_{j \neq i} \mathcal{R}_j = \text{span}\,\{\partial/\partial z_j : j \neq i\}$, $i = 1, \dots, m$, $\mathcal{R}^* = \text{span}\{\partial/\partial z_{m+1,1}, \partial/\partial z_{m+1,2}\}$ and $\Delta^*_{MIX} = \text{span}\{\partial/\partial z_{m+1,2}\}$. Since by assumption (4.7) is locally asymptotically stable at the origin, if

$$\begin{aligned}
u_i &= \alpha_i(z_1, \dots, z_m, z_{m+1,1}, w) + \sum_{j=1}^{m} \beta_{ij}(z_1, \dots, z_m, z_{m+1,1}, w)v_j, \qquad i = 1, \dots, m, \\
\dot{w} &= \delta(z_1, \dots, z_m, z_{m+1,1}, w) + \gamma(z_1, \dots, z_m, z_{m+1,1}, w)v
\end{aligned} \tag{4.21}$$

is a feedback law which solves NLSD for the system

$$\begin{aligned}
\dot{z}_i &= \widetilde{f}_i(z_i) + \widetilde{g}_{ii}(z_i)u_i, \qquad i = 1, \dots, m, \\
\dot{z}_{m+1,1} &= \widetilde{f}_{m+1,1}(z_1, \dots, z_m, z_{m+1,1}) + \sum_{j=1}^{m} \widetilde{g}_{m+1,1j}(z_1, \dots, z_m, z_{m+1,1})v_j, \\
y_i &= h_i(z_i), \qquad i = 1, \dots, m,
\end{aligned} \tag{4.22}$$

it easily follows that also the composite system (3.8)–(4.21) is locally asymptotically stable at the origin of the state space. Moreover, since $\Delta^*_{MIX} \subset \ker\,\{dh\}$, (3.8)–(4.21) is also noninteractive. This proves that, under the assumption that (4.7) is locally asymptotically stable at the origin, if NLSD is solvable for (4.22), then it is solvable also for (3.8). Now, let $z = x$ for simplicity, σ be the natural projection $\sigma(x) = (x_1^T, \dots, x_{m+1,1}^T)^T$ and $\widehat{f}(\sigma(x)) = \sigma_{*x}\widetilde{f}(x)$, $\widehat{g}_i(\sigma(x)) = \sigma_{*x}\widetilde{g}_j(x)$ and $\widehat{h}_i(\sigma(x)) = h_i(x)$ for $i = 1, \dots, m$. If, in addition, we will show that assumptions similar to (A1), (A2), (A3) and (A4) hold also for (4.22), then, without any further doubt, we can assume $\Delta^*_{MIX} = 0$ in a neighbourhood of x_0. For, let $\widehat{\Delta}_{MIX}$ be defined as Δ_{MIX} but with $\widetilde{f}$ and $\widetilde{g}_j$, $j = 1, \dots, m$, replaced by $\widehat{f}$ and $\widehat{g}_j$, $j = 1, \dots, m$, respectively, and $\widehat{\mathcal{R}}_i$ be the maximal local controllability distributions for (4.22) contained in $\bigcap_{j \neq i} \ker \{d\widehat{h}_j\}$. For our purposes, it is sufficient to prove the following lemma.

Lemma 4.6. Let σ, $\widehat{f}, \widehat{g}_i$ and $\widehat{h}_i$, $i = 1, \dots, m$, be as above. Then for all x in a neighbourhood of x_0

$$\sigma_{*x}(\Delta^*_{MIX})(x) = \widehat{\Delta}_{MIX}(\sigma(x)) = 0,$$
$$\sigma_{*x}(\mathcal{R}^*_i + \Delta^*_{MIX})(x) = \widehat{\mathcal{R}}_i(\sigma(x)) = \langle \widehat{f}, \widehat{g}_1, \dots, \widehat{g}_m | \mathrm{span}\{\widehat{g}_i\}\rangle(\sigma(x)), \qquad i = 1, \dots, m.$$

Moreover, (4.22) is strongly accessible at $\sigma(x_0)$ and has vector relative degree at $\sigma(x_0)$. □

Proof. We know that $\mathcal{R}^*_i + \Delta^*_{MIX}$ is spanned by the vector fields of the form either

$$\begin{aligned} &[\widetilde{g}_{j_h}, [\dots, [\widetilde{g}_{j_1}, \widetilde{g}_i] \dots]], \\ &\qquad j_h, \dots, j_1 \in \{0, \dots, m\}\,,\ \widetilde{g}_0(x) = \widetilde{f}(x), \end{aligned} \tag{4.23}$$

or

$$\begin{aligned} &[\widetilde{g}_{j_h}, [\dots, [\widetilde{g}_{j_2}, \widetilde{g}_{j_1}] \dots]], \\ &\qquad j_h, \dots, j_1 \in \{0, \dots, m\}\,,\ \widetilde{g}_0(x) = \widetilde{f}(x)\,, \\ &\qquad j_{k_1} \neq j_{k_2} \text{ and } j_{k_1}, j_{k_2} \neq 0 \text{ for some } k_1\,,\ k_2 \in \{1, \dots, h\}. \end{aligned}$$

Assume that for all x in a neighbourhood of x_0

$$\sigma_{*x}([\widetilde{g}_{j_h}, [\dots, [\widetilde{g}_{j_1}, \widetilde{g}_i] \dots]])(x) = [\widehat{g}_{j_h}, [\dots, [\widehat{g}_{j_1}, \widehat{g}_i] \dots]](\sigma(x))$$

for some $h \geq 1, j_h, \dots, j_1 \in \{0, \dots, m\}$, $\widetilde{g}_0(x) = \widetilde{f}(x)$ and $\widehat{g}_0(\sigma(x)) = \widehat{f}(\sigma(x))$. Thus,

$$\begin{aligned} \sigma_{*x}([\widetilde{g}_{j_{h+1}}, [\dots, [\widetilde{g}_{j_1}, \widetilde{g}_i] \dots]])(x) &= [\sigma_{*x}(\widetilde{g}_{j_{h+1}}), \sigma_{*x}([\widetilde{g}_{j_h}, [\dots, [\widetilde{g}_{j_1}, \widetilde{g}_i] \dots]])](\sigma(x)) = \\ &= [\widehat{g}_{j_h}, [\widehat{g}_{j_{h+1}}, \dots, [\widehat{g}_i, \widehat{g}_i] \dots]](\sigma(x)), \end{aligned}$$

which proves by induction that $\sigma_{*x}(\Delta^*_{MIX})(x) = \widehat{\Delta}_{MIX}(\sigma(x)) = 0$ for all x in a neighbourhood of x_0.

Now, we will show that $\langle \widehat{f}, \widehat{g}_1, \dots, \widehat{g}_m | \mathrm{span}\{\widehat{g}_i\}\rangle(\sigma(x)) = \widehat{\mathcal{R}}_i(\sigma(x))$ for all x in a neighbourhood of x_0. For, note that the vector field $\sigma_*(\tau_i)$, with τ_i a vector field of the form (4.23), is contained in $\langle \widehat{f}, \widehat{g}_1, \dots, \widehat{g}_m | \mathrm{span}\{\widehat{g}_i\}\rangle$. On the other hand, the distribution spanned by the vector fields $\{\sigma_*(V_i)\ :\ V_i \text{ vector field of the form (4.23)}\}$ is locally invariant under $\widehat{f}, \widehat{g}_1, \dots, \widehat{g}_m$ and contains $\widehat{g}_i$. Thus, by definition it contains also $\langle \widehat{f}, \widehat{g}_1, \dots, \widehat{g}_m | \mathrm{span}\{\widehat{g}_i\}\rangle$.

It follows that $\langle \widehat{f}, \widehat{g}_1, \ldots, \widehat{g}_m | \text{span}\{\widehat{g}_i\}\rangle(\sigma(x)) = \sigma_{*x}(\mathcal{R}_i^* + \Delta_{MIX}^*)(x)$ for all x in a neighbourhood of x_0. Moreover, since $\langle \widehat{f}, \widehat{g}_1, \ldots, \widehat{g}_m | \text{span}\{\widehat{g}_i\}\rangle$ has constant dimension in a neighbourhood of $\sigma(x_0)$, it is locally involutive and, thus, it is a local controllability distribution for (4.22). Since $\mathcal{R}_i^* \subset \bigcap_{j \neq i} \mathcal{K}_j$, from proposition 3.1 it follows that $\langle \widehat{f}, \widehat{g}_1, \ldots, \widehat{g}_m | \text{span}\{\widehat{g}_i\}\rangle$ is contained in $\bigcap_{j \neq i} \ker \{d\widehat{h}_j\}$. Moreover, it must be the *maximal* distribution with these properties. Indeed, since Δ_{MIX}^* is locally controlled invariant and is contained in $\bigcap_{j \neq i} \mathcal{K}_j$, we conclude that $\Delta_{MIX}^* \subset \mathcal{V}_i^*$, where $\mathcal{V}_i^*$ is the maximal locally controlled invariant distribution contained in $\bigcap_{j \neq i} \mathcal{K}_j$. It follows from proposition **A**.8 of [40] that $\sigma_*(\mathcal{V}_i^*)$ is the maximal locally controlled invariant distribution contained in $\bigcap_{j \neq i} \ker \{d\widehat{h}_j\}$. Since the feedback law (3.1), for which we obtain (3.8) from (1.1), renders locally invariant the distributions $\mathcal{V}_i^*$, $i = 1, \ldots, m$, under $\widetilde{f}, \widetilde{g}_1, \ldots, \widetilde{g}_m$ (see [52], lemma 6.3.13 for a proof), also the distributions $\sigma_*(\mathcal{V}_i^*)$, $i = 1, \ldots, m$, are locally invariant under $\widehat{f}, \widehat{g}_1, \ldots, \widehat{g}_m$. Thus, the maximal local controllability distribution contained in $\sigma_*(\mathcal{V}_i^*)$ (or, equivalently, in $\bigcap_{j \neq i} \ker \{d\widehat{h}_j\}$) is exactly $\langle \widehat{f}, \widehat{g}_1, \ldots, \widehat{g}_m | \sigma_*(\mathcal{V}_i^*) \cap \text{span}\{\widehat{g}_1, \ldots, \widehat{g}_m\}\rangle$. Our claim follows from the fact that $\sigma_*(\mathcal{V}_i^*) \cap \text{span}\{\widehat{g}_1, \ldots, \widehat{g}_m\} = \text{span}\{\widehat{g}_i\}$. The strong accessibility of (4.22) at $\sigma(x_0)$ can be proved using arguments similar to those above. The fact that (4.22) has vector relative degree at $\sigma(x_0)$ is an easy consequence of assumption (A1) and the fact that $\Delta_{MIX}^* \subset \bigcap_{i=1}^{m} \mathcal{K}_i$.
□

Note that $\widehat{\mathcal{R}}_i$ is *not* necessarily regularly computable at $\sigma(x_0)$. However, as it will be clear later, the crucial fact for our proofs to hold is the fact that there must exist vector fields of the form

$$[\widehat{g}_{j_h}, [\cdots, [\widehat{g}_{j_1}, \widehat{g}_i] \cdots]],$$
$$j_h, \ldots, j_1 \in \{0, \ldots, m\}\ ,\ \widehat{g}_0 = \widehat{f},$$

linearly independent at each point of a neighbourhood of $\sigma(x_0)$ and spanning $\widehat{\mathcal{R}}_i$.

We will proceed according to the following strategy. First, we define an *extended* system

$$\begin{aligned} \dot{x} &= \widetilde{f}(x) + \sum_{j=1}^{m} \widetilde{g}_j(x) u_j, \\ \dot{w} &= u^w, \qquad w \in \mathbb{R}^{n^w}, \\ y_i &= h_i(x), \qquad i = 1, \ldots, m, \end{aligned} \tag{4.24}$$

where $\dot{x} = \widetilde{f}(x) + \sum_{j=1}^{m} \widetilde{g}_j(x) u_j$ is the system (4.22). The extended system has $m + n_w$ inputs and m outputs. The introduction of the *fictitious* inputs u^w allows us to look at the problem of achieving noninteraction and stability for (4.22) via *dynamic* feedback as the problem of achieving noninteraction and stability for (4.24) via *static* feedback laws $u^e = \alpha^e(x^e) + \beta^e(x^e)v^e$ with a particular structure (see section **2**.1).

As a second step, we define for (4.24) m local controllability distributions $\mathcal{R}_i^e$, $i = 1,\dots,m$, which assign at each (x,w) a subspace of $\bigcap_{j\neq i}\mathcal{K}_i(x) \oplus \mathbb{R}^{n^w}$, $i = 1,\dots,m$, respectively. Moreover, these distributions are locally *involutive*, linearly *independent* at each point of a neighbourhood of x_0^e (propositions 4.7 and 4.11 and remark 4.10) and locally *compatible* (proposition 4.12). Once we render $\mathcal{R}_1^e,\dots,\mathcal{R}_m^e$ locally invariant through an invertible static feedback law, we obtain noninteraction. However, the system $\widetilde{\Sigma}^e$, obtained from plugging into (4.24) this invertible static feedback law, is not necessarily locally asymptotically stable in x_0^e. The crucial fact is that now the controllability distributions $\mathcal{R}_1^e,\dots,\mathcal{R}_m^e$, unlike the distributions $\mathcal{R}_1^*,\dots,\mathcal{R}_m^*$, are linearly independent at each x^e in a neighbourhood of x_0^e. In other words, the use of dynamic extension helps enlarge the state space so that to make enough room to "separate" each other the controllability distributions $\mathcal{R}_1^*,\dots,\mathcal{R}_m^*$. Moreover, the definition of $\mathcal{R}_i^e$ is such that all the properties of $\mathcal{R}_i^*$ are inherited by $\mathcal{R}_i^e$. In particular, since each system (4.8) is locally asymptotically stabilizable at the origin via dynamic feedback, also each corresponding control system, described by the vector fields $\widetilde{f}|\mathcal{L}_{x_0^e}^{\mathcal{R}_i^e}$ and $\widetilde{g}_i|\mathcal{L}_{x_0^e}^{\mathcal{R}_i^e}$, is locally asymptotically stabilizable at the origin via dynamic feedback (note that, in the linear case, this property is automatically guaranteed by the fact that $\mathcal{R}_i^e$ is a controllability subspace). This property, together with the fact that (4.22) is strongly accessible at x_0 (lemma 4.6), ensures that $\widetilde{\Sigma}^e$ is locally asymptotically stabilizable at the origin of the state space without destroying the invariance property of the controllability distributions $\mathcal{R}_1^e,\dots,\mathcal{R}_m^e$ (proposition 4.16). This exactly amounts to solving NLSD.

Let us go through the technical details. If

$$w_i = \begin{pmatrix} \lambda_i \\ \mu_i \end{pmatrix}, \quad w = \begin{pmatrix} w_1 \\ \vdots \\ w_m \end{pmatrix},$$

$$\dim \lambda_i = \dim x_i = n_i\,,\ \dim \mu_i = \dim x_{m+1,1} = n_0, \qquad i = 1,\dots,m,$$

$$x^e = \begin{pmatrix} x \\ w \end{pmatrix}, \quad u^e = \begin{pmatrix} u \\ u^w \end{pmatrix},$$

$$f^e(x^e) = \begin{pmatrix} \widetilde{f}(x) \\ 0_{n^w\times 1} \end{pmatrix}, \quad g_i^e(x^e) = \begin{pmatrix} \widetilde{g}_i(x) \\ 0_{n^w\times 1} \end{pmatrix}, \quad g_{m+i}^e(x^e) = \frac{\partial}{\partial w_i}, \qquad i = 1,\dots,m,$$

$$G^e(x^e) = (\,g_1^e(x^e) \quad \dots \quad g_{2m}^e(x^e)\,)\,,\ h_i^e(x^e) = h_i(x), \qquad i = 1,\dots,m,$$

we can rewrite (4.24) as

$$\begin{aligned} \dot{x}^e &= f^e(x^e) + G^e(x^e)u^e, \\ y_i^e &= h_i^e(x^e), \qquad i = 1,\dots,m. \end{aligned} \tag{4.25}$$

Moreover,

$$\mathcal{G}_i^w = \text{span}\,\{g_{m+i}^e\}\ ,\ \mathcal{G}^w = \sum_{i=1}^{m} \mathcal{G}_i^w\ ,\ \mathcal{G}^e = \text{span}\,\{G^e\}, \qquad i = 1, \ldots, m.$$

Note that the dimension of the dynamic extension is $\sum_{i=1}^{m} (n_i + n_0) = n + (m-1)n_0$. In section 2.3 this dimension has been taken equal to $\sum_{i=1}^{m} s_i$, which is less than $\sum_{i=1}^{m} (n_i + n_0)$, since $\mathcal{R}_i^* \subset \mathcal{R}^* + \text{span}\{\partial/\partial x_i\}$. In section 2.5, we shall give conditions under which the dimension of n^w can be reduced to $\sum_{i=1}^{m} s_i$ or, better, to $\sum_{i=2}^{m} \dim\,(\mathcal{R}_i^* \cap \sum_{j<i} \mathcal{R}_j^*)$.

Let $s_i = \dim \mathcal{R}_i^*$ and choose vector fields $X_{i1}, \ldots, X_{is_i}$ of the form (4.23) such that $\mathcal{R}_i^* = \text{span}\{X_{ik} : k = 1, \ldots, s_i\}$ in a neighbourhood of x_0. By direct inspection, it is easy to realize that the vector fields X_{ik}, $k = 1, \ldots, s_i$, have the following structure

$$X_{ik}(x) = \begin{pmatrix} 0_{n_1 \times 1} \\ \vdots \\ 0_{n_{i-1} \times 1} \\ Y_{ik}(x_i) \\ 0_{n_{i+1} \times 1} \\ \vdots \\ Z_{ik}(x) \end{pmatrix}. \tag{4.26}$$

Let us define the following vector fields $X_{ik}^e \in \mathbb{R}^{n+n^w}$

$$X_{ik}^e(x^e) = \begin{pmatrix} X_{ik}(x) \\ 0_{(n_1+n_0)\times 1} \\ \vdots \\ X_{ik}^*(x_i, \mu_i) \\ 0_{(n_{i+1}+n_0)\times 1} \\ \vdots \\ 0_{(n_m+n_0)\times 1} \end{pmatrix}, \qquad i = 1, \ldots, m,\ k = 1, \ldots, s_i, \tag{4.27}$$

where

$$X_{ik}^*(x_i, \mu_i) = \begin{pmatrix} Y_{ik}(x_i) \\ Z_{ik}(x) \end{pmatrix} \Bigg|_{x_j = 0 \text{ for } j \notin \{i,(m+1,1)\}\ ;\ x_{m+1,1} = \mu_i}$$

and, correspondingly, define the following distributions

$$\mathcal{R}_i^e = \text{span}\{X_{ik}^e : 1 \le k \le s_i\}, \qquad i = 1, \ldots, m. \tag{4.28}$$

The key properties of the distributions $\mathcal{R}_1^e, \ldots, \mathcal{R}_m^e$ are summed up in the following proposition.

Proposition 4.7. The distributions $\mathcal{R}_i^e$, $i = 1,\ldots,m$, have constant dimension s_i in a neighbourhood of x_0^e, are linearly independent at each x^e of a neighbourhood of x_0^e, locally controlled invariant for (4.25) and contained, respectively, in $\bigcap_{j\neq i} \ker\{dh_j^e\}$, $i = 1,\ldots,m$.□

Proof. The dimension of $\mathcal{R}_i^e$ follows directly from the fact that the first n components of $X_{ik}^e(x^e)$ are equal to those of $X_{ik}(x)$.

Moreover, the distributions $\mathcal{R}_1^e,\ldots,\mathcal{R}_m^e$ are linearly independent at each x^e in a neighbourhood of x_0^e or, equivalently, $\mathcal{R}_i^e \cap \sum_{j\neq i}\mathcal{R}_j^e = 0$ in a neighbourhood of x_0^e and for $i = 1,\ldots,m$. Indeed, if this were not true, there would exist in each neighbourhood of x_0^e a point $\bar{x}^e$ such that the matrix $(X_{i1}^*(\bar{x}^e) \quad \cdots \quad X_{is_i}^*(\bar{x}^e))$ has rank strictly less than s_i. Clearly, this would contradict the linear independence of the vector fields $X_{i1},\ldots,X_{is_i}$ at each x of a neighbourhood of x_0.

The distributions $\mathcal{R}_1^e,\ldots,\mathcal{R}_m^e$ are also locally controlled invariant for (4.25), since

$$
\begin{aligned}
[f^e,\mathcal{R}_i^e] &\subset [\tilde{f},\mathcal{R}_i^*] + \mathcal{G}^w \subset \mathcal{R}_i^e + \mathcal{G}^w,\\
[g_j^e,\mathcal{R}_i^e] &\subset [\tilde{g}_j,\mathcal{R}_i^*] + \mathcal{G}^w \subset \mathcal{R}_i^e + \mathcal{G}^w, \qquad j = 1,\ldots,m,\\
[g_{m+j}^e,\mathcal{R}_i^e] &\subset [\frac{\partial}{\partial w_j},\mathcal{R}_i^*] + \mathcal{G}^w \subset \mathcal{R}_i^e + \mathcal{G}^w, \qquad j = 1,\ldots,m
\end{aligned}
$$

(here, the distributions $\mathcal{R}_i^*$, $[\tilde{f},\mathcal{R}_i^*]$ and $[\tilde{g}_j,\mathcal{R}_i^*]$ are thought of as defined in a neighbourhood of x_0^e). Moreover, they are contained in $\bigcap_{j\neq i} \ker\{dh_j^e\}$, since $\mathcal{R}_i^* \subset \bigcap_{j\neq i}\ker\{dh_j\}$.

Finally, we prove that $\mathcal{R}_i^e$ is locally involutive. For, since $\mathcal{R}_i^*$ is locally involutive, we have for all x in a neighbourhood of x_0

$$
[X_{ik},X_{ih}](x) = \sum_{t=1}^{s_i} c_{ikht}(x)X_{it}(x)\,, \tag{4.29}
$$

where the c_{ikht}'s are unique smooth real-valued functions. We claim that these functions depend only on x_i. Indeed, since $\Delta_{MIX}^* = 0$ implies that $[X_{ik},X_{jr}](x) = 0$ for $j \neq i$, for all vector fields X_{ik} of the form (4.23) and for all x in a neighbourhood of x_0, from (4.29) we obtain

$$
0 = [[X_{ik},X_{ih}],X_{jr}](x) = \sum_{t=1}^{s_i}(-L_{X_{jr}}c_{ikht}(x))X_{it}(x)\,, \qquad j \neq i,
$$

for all x in a neighbourhood of x_0. Since the vector fields $X_{i1},\ldots,X_{is_i}$ are linearly independent at each x of a neigbourhood of x_0, it follows that $L_{X_{jr}}c_{ikht}(x) = 0$ for $j \neq i$ and for all x in a neighbourhood of x_0, which, together with the fact that $\sum_{j\neq i}\mathcal{R}_j^* = \text{span}\{\partial/\partial x_j :$

$j \neq i\}$, proves our claim. Note that $[X_{ik}, X_{ih}](x)$ is of the form

$$[X_{ik}, X_{ih}](x) = \begin{pmatrix} 0_{n_1 \times 1} \\ \vdots \\ 0_{n_{i-1} \times 1} \\ Y_{ikh}(x_i) \\ 0_{n_{i+1} \times 1} \\ \vdots \\ Z_{ikh}(x) \end{pmatrix}.$$

If we define

$$[X_{ik}, X_{ih}]^*(x_i, \mu_i) = \begin{pmatrix} Y_{ikh}(x_i) \\ Z_{ikh}(x) \end{pmatrix} \Big|_{x_j=0 \text{ for } j \notin \{i,(m+1,1)\}\ ;\ x_{m+1,1}=\mu_i}$$

it can be easily seen by straightforward computations that

$$[X_{ik}, X_{ih}]^*(x_i, \mu_i) = [X_{ik}^*, X_{ih}^*](x_i, \mu_i).$$

This, together with (4.29), implies that $[X_{ik}, X_{ih}]^*(x_i, \mu_i) = \sum_{t=1}^{s_i} c_{ikht}(x_i) X_{it}^*(x_i, \mu_i)$. Thus, for all x^e in a neighbourhood of x_0^e

$$[X_{ik}^e, X_{ih}^e](x^e) = \begin{pmatrix} [X_{ik}, X_{ih}](x) \\ 0_{(n_1+n_0)\times 1} \\ \vdots \\ 0_{(n_{i-1}+n_0)\times 1} \\ [X_{ik}^*, X_{ih}^*](x_i, \mu_i) \\ 0_{(n_{i+1}+n_0)\times 1} \\ \vdots \\ 0_{(n_m+n_0)\times 1} \end{pmatrix} = \sum_{t=1}^{s_i} c_{ikht}(x_i) X_{it}^e(x^e),$$

which proves that $\mathcal{R}_i^e$ is locally involutive. □

Remark 4.8. In the case of linear systems, since the vector fields X_{ik}^e are constant, the distribution $\mathcal{R}_i^e$ is trivially involutive. □

Remark 4.9. The distribution $\mathcal{R}_i^e$ does not depend on the choice of the vector fields X_{ik}, $k = 1, \ldots, s_i$. Indeed, let X_{ik}, $k = 1, \ldots, s_i$, and X'_{ik}, $k = 1, \ldots, s_i$, be two such choices. Moreover, let X_{ik}^e, $k = 1, \ldots, s_i$, and $(X'_{ik})^e$, $k = 1, \ldots, s_i$, be the corresponding vector fields defined as in (4.27) and $\mathcal{R}_i^e$ and $(\mathcal{R}'_i)^e$ be the corresponding distributions defined as in (4.28). We will prove first that $\mathcal{R}_i^e \subset (\mathcal{R}'_i)^e$. Similarly to (4.29), we obtain $X_{ik}(x) = \sum_{r=1}^{s_i} c_{ikr}(x) X'_{ir}(x)$, where the c_{ikr}'s are unique smooth functions of their arguments and depend only on x_i. It follows that $X^*_{ik}(x_i,\mu_i) = \sum_{r=1}^{s_i} c_{ikr}(x_i)(X'_{ir})^*(x_i,\mu_i)$, where $(X'_{ir})^*(x_i,\mu_i)$ is defined as in (4.27) but with $X_{ik}(x)$ replaced by $X'_{ik}(x)$. As a consequence, $X^e_{ik}(x^e) = \sum_{r=1}^{s_i} c_{ikr}(x_i)(X'_{ir})^e(x^e)$, which implies $\mathcal{R}^e_i \subset (\mathcal{R}'_i)^e$. Similarly, we prove the opposite inclusion. The fact that $\mathcal{R}^e_i$ does not depend on the choice of X_{ik}, $k=1,\ldots,s_i$, allows us to assume $X_{i1}(x) = \widetilde{g}_i(x)$, $i=1,\ldots,m$. □

Remark 4.10. It can be shown that, if $\mathcal{R}^*_i$ is regularly computable at x_0 (see remarks immediately after the proof of lemma 4.6), $\mathcal{R}^e_i$ is a local controllability distribution for (4.25). Indeed, since $\mathcal{R}^*_i$ is regularly computable at x_0, it follows that for the sequence

$$\begin{aligned}
\mathcal{R}_{i0} &= \mathcal{R}^*_i \cap \mathcal{G} = \operatorname{span}\{\widetilde{g}_i\},\\
\mathcal{R}_{ik} &= \mathcal{R}^*_i \cap ([\widetilde{f},\mathcal{R}_{i,k-1}] + \sum_{j=1}^{m}[\widetilde{g}_j,\mathcal{R}_{i,k-1}] + \mathcal{G}) + \mathcal{R}_{i,k-1} =\\
&= [\widetilde{f},\mathcal{R}_{i,k-1}] + \sum_{j=1}^{m}[\widetilde{g}_j,\mathcal{R}_{i,k-1}] + \mathcal{R}_{i,k-1},
\end{aligned}$$

there exists k^*_i such that $\mathcal{R}_{ik^*_i} = \mathcal{R}_{i,k^*_i+1} = \mathcal{R}^*_i$ and each distribution $\mathcal{R}_{ik}$ has constant dimension in a neighbourhood of x_0. Consider now the sequence

$$\begin{aligned}
\mathcal{R}^e_{i0} &= \mathcal{R}^e_i \cap \mathcal{G}^e,\\
\mathcal{R}^e_{ik} &= \mathcal{R}^e_i \cap ([f^e,\mathcal{R}^e_{i,k-1}] + \sum_{j=1}^{2m}[g^e_j,\mathcal{R}^e_{i,k-1}] + \mathcal{G}^e) + \mathcal{R}^e_{i,k-1}.
\end{aligned}$$

Since $\mathcal{R}^e_i$ is locally controlled invariant for (4.25) and involutive, it is also a local controllability distribution for (4.25) if we show that for the above sequence there exists h^*_i such that $\mathcal{R}^e_{ih^*_i} = \mathcal{R}^e_{i,h^*_i+1} = \mathcal{R}^e_i$ (see [55], lemma 6.4.4). For, note that $\mathcal{R}^e_i \cap \mathcal{G}^e = \operatorname{span}\{X^e_{i1}\}$. Indeed, by construction $\mathcal{R}^e_i \cap \mathcal{G}^w = 0$ in a neighbourhood of x^e_0, since otherwise the vector fields X_{ik}, $k=1,\ldots,s_i$, would not be linearly independent at each x in a neighbourhood of x_0. This, together with $\mathcal{R}^*_i \cap \mathcal{G} = \operatorname{span}\{X_{i1}\}$ (see remark 4.9), proves that $\mathcal{R}^e_i \cap \mathcal{G}^e = \operatorname{span}\{X^e_{i1}\}$.

The distributions $\mathcal{R}_{ik}, k \geq 0$, have constant dimension in a neighbourhood of x_0 and, thus, are spanned by the vector fields of the form (4.23). Moreover, $\mathcal{R}^e_{i0} = \mathcal{R}^e_i \cap \mathcal{G}^e = \operatorname{span}\{X^e_{i1}\}$ and $\dim \mathcal{R}^e_{i0}(x^e) = 1 = \dim \mathcal{R}_{i0}(x)$. Suppose that for some $k > 0$ the distributions $\mathcal{R}_{ik}$ and $\mathcal{R}^e_{ik}$ have constant dimension equal to s_{ik} in a neighbourhood of x_0 and x^e_0, respectively, and $\mathcal{R}^e_{ik} = \operatorname{span}\{X^e_{i1},\ldots,X^e_{is_{ik}}\}$ in a neighbourhood of x^e_0, where $X^e_{i1},\ldots,X^e_{is_{ik}}$ are defined as in (4.27) with $X_{i1},\ldots,X_{is_{ik}}$ vector fields of the form (4.23) such that $\mathcal{R}_{ik} = \text{span}\{X_{i1}, \ldots, X_{s_{ik}}\}$ in a neighbourhood of x_0. By induction hypothesis,

$$\begin{aligned}\mathcal{R}^e_{i,k+1} &= \mathcal{R}^e_i \cap ([f^e, \mathcal{R}^e_{ik}] + \sum_{j=1}^{2m} [g^e_j, \mathcal{R}^e_{ik}] + \mathcal{G}^e) + \mathcal{R}^e_{ik} = \\ &= \mathcal{R}^e_i \cap (\text{span}\{[f^e, X^e_{ih}], [g^e_j, X^e_{ih}] : j = 1, \ldots, m, h = 1, \ldots, s_{ik}\} + \mathcal{G}^e) + \mathcal{R}^e_{ik} = \\ &= \mathcal{R}^e_i \cap (\text{span}\{[\widetilde{f}, X_{ih}]^e, [\widetilde{g}_j, X_{ih}]^e : j = 1, \ldots, m, h = 1, \ldots, s_{ik}\} + \mathcal{G}^e) + \mathcal{R}^e_{ik},\end{aligned} \tag{4.30}$$

where $[\widetilde{f}, X_{ik}]^e(x^e)$ and $[\widetilde{g}_j, X_{ik}]^e(x^e)$ are defined as in (4.27) but replacing $X_{ik}(x)$ with $[\widetilde{f}, X_{ik}](x)$ and $[\widetilde{g}_j, X_{ik}](x)$, respectively.Since by construction $[\widetilde{f}, X_{ik}]^e \in \mathcal{R}^e_i$, $[\widetilde{g}_j, X_{ik}]^e \in \mathcal{R}^e_i$ and $\mathcal{R}^e_i \cap \mathcal{G}^w = 0$, it follows from (4.30) that

$$\mathcal{R}^e_{i,k+1} = \text{span}\{[\widetilde{f}, X_{ih}]^e, [\widetilde{g}_j, X_{ih}]^e : j = 1, \ldots, m, h = 1, \ldots, s_{ik}\} + \mathcal{R}^e_{ik}.$$

By construction, there exist $s_{i,k+1} \geq s_{ik}$ vector fields in the set $\{[\widetilde{f}, X_{ih}]^e, [\widetilde{g}_j, X_{ih}]^e, X^e_{ih} : j = 1, \ldots, m,\ h = 1, \ldots, s_{ik}\}$, which are linearly independent for all x^e in a neighbourhood of x^e_0, if and only if the corresponding projected vector fields in the set $\{[\widetilde{f}, X_{ih}], [\widetilde{g}_j, X_{ih}], X_{ih} : j = 1, \ldots, m,\ h = 1, \ldots, s_{ik}\}$ are linearly independent for all x in a neighbourhood of x_0.

Thus, $\mathcal{R}_{i,k+1}$ and $\mathcal{R}^e_{i,k+1}$ have constant dimension equal to $s_{i,k+1} \geq s_{ik}$ in a neighbourhood of x_0 and x^e_0, respectively, and $\mathcal{R}^e_{i,k+1} = \text{span}\{X^e_{ih} : h = 1, \ldots, s_{i,k+1}\}$ in a neighbourhood of x^e_0, where X^e_{ih} are defined as in (4.27) with $X_{i1}, \ldots, X_{is_{i,k+1}}$ vector fields of the form (4.23) such that $\mathcal{R}_{i,k+1} = \text{span}\{X_{ih} : h = 1, \ldots, s_{i,k+1}\}$ in a neighbourhood of x_0. This completes by induction our proof.

If the distributions $\sum_{j \neq i} \mathcal{R}^*_j$ are regularly computable at x_0, then, by an induction argument similar to the one above, it can be shown that also the distributions $\sum_{j \neq i} \mathcal{R}^e_j$, $i = 1, \ldots, m$, are local controllability distributions for (4.25). □

The following proposition is a consequence of the structure of the vector fields X^e_{ik}, $k = 1, \ldots, s_i$.

Proposition 4.11. The distributions $\sum_{j \in I} \mathcal{R}^e_j$, $I \subset \{1, \ldots, m\}$, *are locally controlled invariant distributions for (4.25).* □

Proof. From $\Delta^*_{MIX} = 0$ it follows that $[X_{ik}, X_{jh}](x) = 0$ for $j \neq i$ and for all x in a neighbourhood of x_0. By straightforward computations, we obtain $[X^e_{ik}, X^e_{jh}](x^e) = 0$ for $j \neq i$ and for all x^e in a neighbourhood of x^e_0. This implies that the distributions $\sum_{j \in I} \mathcal{R}^e_j$, $I \subset \{1, \ldots, m\}$, are locally involutive. Moreover, they are locally controlled invariant for (4.25), since each distribution $\mathcal{R}^e_i$ is. □

As we will show in a moment, the distributions $\mathcal{R}_1^e, \ldots, \mathcal{R}_m^e$ are locally *compatible*, i.e. there exists a feedback law $u^e = \alpha^e(x^e) + \sum_{j=1}^{2m} \beta_j^e(x^e) v_j^e$, defined in a neighbourhood of x_0^e and such that $[f^e + G^e\alpha^e, \mathcal{R}_i^e] \subset \mathcal{R}_i^e$ and $[G^e\beta_j^e, \mathcal{R}_i^e] \subset \mathcal{R}_i^e$ for $i = 1, \ldots, m$ and $j = 1, \ldots, 2m$. Moreover, this feedback law is *invertible*.

Proposition 4.12. There exists an invertible feedback law $u^e = \alpha^e(x^e) + \sum_{j=1}^{2m} \beta_j^e(x^e) v_j^e$, $\alpha^e(x_0^e) = 0$, *defined in a neighbourhood of* x_0^e *and such that*

$$[f^e + G^e\alpha^e, \mathcal{R}_i^e] \subset \mathcal{R}_i^e, \tag{4.31}$$

$$[G^e\beta_j^e, \mathcal{R}_i^e] \subset \mathcal{R}_i^e, \qquad j = 1, \ldots, 2m, \tag{4.32}$$

and $G^e\beta_i^e = X_{i1}^e$ *for* $i = 1, \ldots, m$. □

Before proving the above proposition, we need the following result.

Lemma 4.13. Let U *and* V *be, respectively, open sets of* $\mathbb{R}^n$ *and* $\mathbb{R}^q$ *and* S *be an involutive distribution, defined on* U *and spanned by some vector fields* $X_1, \ldots, X_s$, *linearly independent at each* $x \in U$. *Let* c_{hkt} *be smooth real-valued functions, defined on* U, *such that* $[X_k, X_h](x) = \sum_{t=1}^{s} c_{hkt}(x) X_t(x)$. *Moreover, let* $\Gamma_1, \ldots, \Gamma_s$ *be smooth functions* $\Gamma_i : U \to \mathbb{R}^{q\times q}$. *Given the set of partial differential equations*

$$L_{X_k}\alpha(x) = \Gamma_k(x)\alpha(x), \qquad k = 1, \ldots, s, \tag{4.33}$$

with $\alpha : U \to V$, *and given* $x_0 \in U$, *there exists a neighbourhood* $U_0 \subset U$ *of* x_0 *and a smooth function* $\alpha : U_0 \to V$, *which satisfies (4.33) if only if*

$$\Gamma_k(x)\Gamma_h(x) - \Gamma_h(x)\Gamma_k(x) + L_{X_h}\Gamma_k(x) - L_{X_k}\Gamma_h(x) = \sum_{t=1}^{s} c_{hkt}(x)\Gamma_t(x)$$

for all $x \in U$. *Moreover, given a submanifold* $\mathcal{M} \subset \mathbb{R}^n$, *complementary to* $\mathcal{L}_{x_0}^S$ *and passing through* x_0, $\alpha(x)$ *can be uniquely chosen in such a way to coincide on* $\mathcal{M} \cap U$ *with a given function* $\bar{\alpha}(x)$. □

Proof. We can follow the proof of theorem 6.2.3 ([55]), if we define Δ as the distribution which assigns at each $(x, y) \in U \times V$ the subspace generated by the vectors

$$\begin{pmatrix} X_1(x) \\ \Gamma_1(x)y \end{pmatrix}, \ldots, \begin{pmatrix} X_s(x) \\ \Gamma_s(x)y \end{pmatrix}.$$

□

Proof (of proposition 4.12). Let $[\widetilde{f}, X_{ik}]^e(x^e)$ be defined as in (4.27) with $X_{ik}(x)$ replaced by $[\widetilde{f}, X_{ik}](x)$. Since $[\widetilde{f}, X_{ik}] \in \mathcal{R}_i^*$, we can write

$$[f^e, X_{ik}^e](x^e) = [\widetilde{f}, X_{ik}]^e(x^e) + g_{m+i}^e(x^e)c_{ik}(x^e), \tag{4.34}$$

where the c_{ik}'s are unique smooth functions $c_{ik} : U^e \to \mathbb{R}^{n_i+n_0}$, U^e a neighbourhood of x_0^e. By using the same arguments as in proposition 4.7, it can be easily shown that $[\widetilde{f}, X_{ik}]^e \in \mathcal{R}_i^e$. Since for any smooth functions $\alpha_{m+1}(x^e), \ldots, \alpha_{2m}(x^e)$ we have

$$[f^e + \sum_{j=m+1}^{2m} g_j^e\alpha_j^e, X_{ik}^e](x^e) = g_{m+i}^e(x^e)(c_{ik}(x^e) - L_{X_{ik}^e}\alpha_i^e(x^e) + \gamma_{ik}(x^e)\alpha_i^e(x^e)) +$$

$$- \sum_{j \neq i} g_{m+j}^e(x^e)L_{X_{ik}^e}\alpha_j^e(x^e) + [\widetilde{f}, X_{ik}]^e(x^e),$$

where $\gamma_{ik}(x^e)$ is the $(n_i + n_0) \times (n_i + n_0)$ matrix, defined in a neighbourhood of x_0^e and with smooth entries, such that

$$[g_{m+i}^e, X_{ik}^e](x^e) = g_{m+i}^e(x^e)\gamma_{ik}(x^e), \tag{4.35}$$

in order to satisfy (4.31) it suffices to solve the following set of partial differential equations

$$L_{X_{jh}^e}\alpha_i^e(x^e) = 0, \qquad j \neq i, \tag{4.36}$$

$$L_{X_{ik}^e}\alpha_i^e(x^e) = c_{ik}(x^e) + \gamma_{ik}(x^e)\alpha_i(x^e), \tag{4.37}$$

or, equivalently,

$$L_{X_{jh}^e}\begin{pmatrix} \alpha_i^e(x^e) \\ \alpha_{2m+1}^e(x^e) \end{pmatrix} = \Gamma_{jh}(x^e)\begin{pmatrix} \alpha_i^e(x^e) \\ \alpha_{2m+1}^e(x^e) \end{pmatrix}, \tag{4.38}$$

where

$$\Gamma_{jh}(x^e) = \begin{cases} \begin{pmatrix} \gamma_{ih}(x^e) & c_{ih}(x^e) \\ 0 & 0 \end{pmatrix} & \text{if } j = i\ , \\ \begin{pmatrix} 0 & 0 \\ 0 & 0 \end{pmatrix} & \text{otherwise}\ , \end{cases} \tag{4.39}$$

and $\alpha_{2m+1}^e(x^e) \equiv 1$ for all x^e in a neighbourhood of x_0^e. The *fictitious* function α_{2m+1}^e has been introduced in order to write the set of partial differential equations (4.36) and (4.37) in the form (4.38). From lemma 4.13 with $\mathcal{S} = \sum_{j=i}^{m} \mathcal{R}_j^e$, it follows that (4.38) is solvable if and only if

$$\Gamma_{ik}(x^e)\Gamma_{jh}(x^e) - \Gamma_{jh}(x^e)\Gamma_{ik}(x^e) + L_{X_{jh}}\Gamma_{ik}(x^e) - L_{X_{ik}}\Gamma_{jh}(x^e) =$$

$$= \begin{cases} \sum_{t=1}^{s} \nu_{ihkt}(x^e)\Gamma_{it}(x^e) & \text{if } j = i\ , \\ \begin{pmatrix} 0 & 0 \\ 0 & 0 \end{pmatrix} & \text{else}\ , \end{cases} \tag{4.40}$$

where the ν_{ihkt}'s are the smooth real-valued functions, defined in a neighbourhood of x_0^e and such that $[X_{ih}, X_{ik}](x^e) = \sum_{t=1}^{s_i} \nu_{ihkt}(x^e) X_{it}(x^e)$. Using (4.39), (4.40) boils down to

$$L_{X_{jh}^e} \gamma_{ik}(x^e) = 0, \qquad j \neq i, \tag{4.41}$$

$$L_{X_{jh}^e} c_{ik}(x^e) = 0, \qquad j \neq i, \tag{4.42}$$

and

$$\sum_{t=1}^{s_i} \nu_{ihkt} \gamma_{it}(x^e) = \gamma_{ik}(x^e)\gamma_{ih}(x^e) - \gamma_{ih}(x^e)\gamma_{ik}(x^e) + L_{X_{ih}^e} \gamma_{ik}(x^e) - L_{X_{ik}^e} \gamma_{ih}(x^e), \tag{4.43}$$

$$\sum_{t=1}^{s_i} \nu_{ihkt}(x^e) c_{it}(x^e) = \gamma_{ik}(x^e) c_{ih}(x^e) - \gamma_{ih}(x^e) c_{ik}(x^e) + L_{X_{ih}^e} c_{ik}(x^e) - L_{X_{ik}^e} c_{ih}(x^e). \tag{4.44}$$

Note that γ_{ik} and c_{ik}, since functions of x_i and μ_i only, clearly satisfy (4.41) and (4.42). On the other hand, the equalities (4.43) and (4.44) can be deduced from the following Jacobi identities

$$- [[f^e, X_{ik}^e], X_{ih}^e](x^e) + [[f^e, X_{ih}^e], X_{ik}^e](x^e) - [f^e, [X_{ih}^e, X_{ik}^e]](x^e) = 0, \tag{4.45}$$

$$- [[g_{m+q}^e, X_{ik}^e], X_{ih}^e](x^e) + [[g_{m+q}^e, X_{ih}^e], X_{ik}^e](x^e) - [g_{m+q}^e, [X_{ih}^e, X_{ik}^e]](x^e) = 0. \tag{4.46}$$

for $i, q = 1, \dots, m$ and for all x^e in a neighbourhood of x_0^e. Indeed, from (4.45) and (4.35) we obtain

$$\begin{aligned}
&- [[f^e, X_{ik}^e], X_{ih}^e](x^e) + [[f^e, X_{ih}^e], X_{ik}^e](x^e) - [f^e, [X_{ih}^e, X_{ik}^e]](x^e) = -[[\tilde{f}, X_{ik}]^e, X_{ih}^e](x^e) + \\
&- g_{m+i}^e(x^e)(\gamma_{ih}(x^e) c_{ik}(x^e) - L_{X_{ih}^e} c_{ik}(x^e)) + [[\tilde{f}, X_{ih}]^e, X_{ik}^e](x^e) + \\
&+ g_{m+i}^e(x^e)(\gamma_{ih}(x^e) c_{ih}(x^e) - L_{X_{ik}^e} c_{ih}(x^e)) + \sum_{t=1}^{s_i} (L_{f^e} \nu_{ihkt}(x^e)) X_{it}^e(x^e) + \\
&- \sum_{t=1}^{s_i} \nu_{ihkt}(x^e) [\tilde{f}, X_{it}]^e(x^e) - \sum_{t=1}^{s_i} \nu_{ihkt}(x^e) g_{m+t}^e(x^e) c_{it}(x^e),
\end{aligned}$$

for all x^e in a neighbourhood of x_0^e, which, since $-[[\tilde{f}, X_{ik}]^e, X_{ih}^e] + [[\tilde{f}, X_{ih}]^e, X_{ik}^e]$ and $[\tilde{f}, X_{it}]^e$ are vector fields of $\mathcal{R}_i^e$ and $\mathcal{R}_i^e \cap \mathcal{G}_i^w = 0$, implies (4.44). Similarly, (4.43) follows from (4.46) with $i = q$. Moreover, from lemma 4.13, a smooth real-valued function $\alpha_i^e(x)$, which satisfies (4.36) and (4.37), can be also chosen in such a way that $\alpha_i^e(x_0^e) = 0$.

Reasoning as above, we can also show that there exist $(n_i + n_0) \times (n_i + n_0)$ matrices $\beta_i^1(x^e)$, $i = 1, \ldots, m$, defined in a neighbourhood of x_0^e and with smooth entries, such that

$$L_{X_{ik}^e} \beta_i^1(x^e) = \gamma_{ik}(x^e)\beta_i^1(x^e), \tag{4.47}$$

$$L_{X_{jh}^e} \beta_i^1(x^e) = 0, \qquad j \neq i, \tag{4.48}$$

and $\beta_i^1(x^e) = I_{(n_i+n_0)\times(n_i+n_0)}$. With this choice, $\beta_i^1(x^e)$ is invertible for all x^e in a neighbourhood of x_0^e, since it is solution of a set of partial differential equations. But, if β_i^1 satisfies (4.47) and (4.48), then it satisfies also

$$[g_{m+i}^e \beta_i^1, X_{jh}^e](x^e) = 0$$

for $i, j = 1, \ldots, m$ and for all x^e in a neighbourhood of x_0^e, which implies $[g_{m+i}^e \beta_i^1, \mathcal{R}_j^e] \subset \mathcal{R}_j^e$ for $j, i = 1, \ldots, m$ and for all x^e in a neighbourhood of x_0^e. Finally, if we define $\beta_i^e(x^e)$, $i = 1, \ldots, m$, in such a way that

$$G^e \beta_i^e(x^e) = X_{i1}^e(x^e),$$

clearly it follows that $[g^e \beta_i^e, \mathcal{R}_j^e] \subset \mathcal{R}_j^e$ for $i, j = 1, \ldots, m$. The feedback law $u^e = \alpha^e(x^e) + \sum_{j=1}^{2m} \beta_j^e(x^e) v_j^e$, with

$$\alpha^e(x^e) = \begin{pmatrix} 0_{m\times 1} \\ \alpha_1^e(x) \\ \vdots \\ \alpha_m^e(x^e) \end{pmatrix},$$

$$\beta_j^e(x^e) = \begin{pmatrix} 0_{m\times(n_j+n_0)} \\ 0_{(n_1+n_0)\times(n_j+n_0)} \\ \vdots \\ 0_{(n_{j-1}+n_0)\times(n_j+n_0)} \\ \beta_j^1(x^e) \\ 0_{(n_{j+1}+n_0)\times(n_j+n_0)} \\ \vdots \\ 0_{(n_m+n_0)\times(n_j+n_0)} \end{pmatrix}, \qquad j = m+1, \ldots, 2m,$$

and β_j^e, $j = 1, \ldots, m$, chosen as above, is an invertible feedback law satisfying (4.31) and (4.32) and $G^e\beta_i(x^e) = X_{i1}^e(x^e)$ for $i = 1, \ldots, m$ and for all x^e in a neighbourhood of x_0. □

Remark 4.14. From the proof of proposition 4.12, it follows that to find the feedback law $u^e = \alpha^e(x^e) + \sum_{j=1}^{2m} \beta_j^e(x^e) v_j^e$ we have to solve a set of partial differential equations. However, we can avoid solving at least (4.36) and (4.37). Indeed, as it can be directly checked, a very simple expression of $\alpha_i^e(x^e)$ is given by

$$\alpha_i^e(x^e) = \begin{pmatrix} \tilde{f}_i(x_i) \\ \tilde{f}_{m+1,1}(0,\ldots,0,x_i,0,\ldots,0,x_{m+1,1}) \end{pmatrix} \Big|_{x_j=0 \text{ for } j\notin\{i,(m+1,1)\}\ ;\ x_{m+1,1}=\mu_i}$$

Moreover, let us denote by $\bar{g}_{m+i}^e(x^e)$ the last n_0 columns of $g_{m+i}^e(x^e)$ and by $\bar{\gamma}_{ik}(x^e)$ the $n_0 \times n_0$ matrix, defined in a neighbourhood of x_0^e and with smooth entries, such that

$$[\bar{g}_{m+i}^e, X_{ik}^e](x^e) = \bar{g}_{m+i}^e(x^e)\bar{\gamma}_{ik}(x^e).$$

Moreover, let $\beta_i^2(x^e)$ be the $n_0 \times n_0$ matrix, defined in a neighbourhood of x_0^e and with smooth entries, such that

$$L_{X_{ik}^e}\beta_i^2(x^e) = \bar{\gamma}_{ik}(x^e)\beta_i^2(x^e),$$
$$L_{X_{jh}^e}\beta_i^2(x^e) = 0, \qquad j \neq i,$$

and $\beta_i^2(x^e) = I_{n_0\times n_0}$. As it can be easily checked, the matrix $\beta_i^1(x^e)$, which satisfies (4.47) and (4.48), can be chosen with the following form

$$\beta_i^1(x^e) = \begin{pmatrix} I_{n_i\times n_i} & 0_{n_i\times n_0} \\ 0_{n_0\times n_i} & \beta_i^2(x^e) \end{pmatrix}.$$

Thus, the number of partial differential equations to be solved is reduced to $n_0^2(\sum_{j=1}^{m} s_j)$. In chapter **3**, a method, based on the theory of parallel transport of vector fields, will be used to avoid solving partial differential equations. This approach allows also to obtain feedback laws satisfying (4.31) and (4.32) and defined on $\mathbb{R}^n$. □

Remark 4.15. If $u^e = \alpha^e(x^e) + \sum_{j=1}^{2m} \beta_j^e(x^e)v_j^e$ is chosen as in the proof of proposition 4.12, then we have

$$\begin{aligned} \tilde{f}^e(x^e) &= \begin{pmatrix} \tilde{f}(x) \\ 0_{n^w\times 1} \end{pmatrix} + \bar{X}(x^e), \qquad \bar{X} \in \mathcal{G}^w, \\ \tilde{g}_j^e(x^e) &= \begin{pmatrix} \tilde{g}_j(x) \\ 0_{n^w\times 1} \end{pmatrix} + \bar{Y}(x^e), \qquad \bar{Y} \in \mathcal{G}^w,\ j = 1,\ldots,m. \end{aligned} \tag{4.49}$$

As a consequence

$$\pi_{*x^e}(\tilde{f}^e(x^e)) = \tilde{f}(\pi(x^e)),$$
$$\pi_{*x^e}(\tilde{g}_j^e(x^e)) = \tilde{g}_j(\pi(x^e)), \qquad j = 1,\ldots,m,$$

for all x^e in a neighbourhood of x_0^e. □

Let $\widetilde{f}^e = f + G^e\alpha^e$ and $\widetilde{g}_j^e = G^e\beta_j^e$, $j = 1, \ldots, 2m$, be as in the proof of proposition 4.12. We have in a neighbourhood of x_0^e

$$
\begin{aligned}
&[\widetilde{f}^e, \mathcal{R}_i^e] \subset \mathcal{R}_i^e, \\
&[\widetilde{g}_j^e, \mathcal{R}_i^e] \subset \mathcal{R}_i^e, \quad j = 1, \ldots, 2m, \\
&\text{span}\{\widetilde{g}_i^e\} = \mathcal{R}_i^e \cap \mathcal{G}^e, \\
&\mathcal{R}_i^e \subset \bigcap_{j \neq i} \mathcal{K}_j^e.
\end{aligned}
\tag{4.50}
$$

More strongly, $\mathcal{R}_i^e = \langle \widetilde{f}^e, \widetilde{g}_1^e, \ldots, \widetilde{g}_m^e | \text{span}\{\widetilde{g}_i^e\}\rangle$ in a neighbourhood of x_0^e. Indeed, since $\mathcal{R}_i^e$ is a local controllability distribution for (4.25), from (4.50) it follows that

$$
\mathcal{R}_i^e = \langle \widetilde{f}^e, \widetilde{g}_1^e, \ldots, \widetilde{g}_{2m}^e | \text{span}\{\widetilde{g}_i^e\}\rangle
$$

in a neighbourhood of x_0^e. But, as it results from the proof of proposition 4.12, $[\widetilde{g}_j^e, X_{ik}^e](x^e) = 0$ for $j = m+1, \ldots, 2m$, for all x^e in a neighbourhood of x_0^e and for *any* vector field X_{ik}^e of the form (4.23), which proves our thesis. As a consequence, the system

$$
\begin{aligned}
\dot{x}^e &= \widetilde{f}^e(x^e) + \sum_{j=1}^{2m} \widetilde{g}_j^e(x^e)v_j^e, \\
y_i &= h_i^e(x^e), \qquad i = 1, \ldots, m,
\end{aligned}
\tag{4.51}
$$

is noninteractive and has vector relative degree at x_0^e when $v_j^e = 0$ for $j = m+1, \ldots, 2m$. Now, we will show that (4.51) can be locally asymptotically stabilized at the origin via dynamic feedback, without destroying the above properties. As expected, this proves that the local asymptotic stability of (4.7) and the local asymptotic stabilizability of each system (4.8) is sufficient for solving NLSD.

Proposition 4.16. There exists a feedback law

$$
\begin{aligned}
v_i^e &= \varphi_i^e(x^e, \eta_i^e) + v_i, \qquad i = 1, \ldots, m, \\
\dot{\eta}_i^e &= \psi_i^e(x^e, \eta_i^e), \qquad i = 1, \ldots, m, \\
v_i^e &= \varphi_i^e(x^e), \qquad i = m+1, \ldots, 2m,
\end{aligned}
\tag{4.52}
$$

such that the system (4.51)-(4.52) resulting from (4.51), is locally asymptotically stable at the origin of the state space, noninteractive and has vector relative degree at the origin of the state space. □

Proof. First, there exists a change of coordinates $z^e(x^e) = \left((z_1^e)^T \;\cdots\; (z_{m+1}^e)^T\right)^T$, $z^e(x_0^e) = 0$, defined in a neighbourhood of x_0^e and such that (4.51) is expressed in these coordinates by

$$\begin{aligned}
\dot{z}_i^e &= \tilde{f}_i^e(z_i^e, z_{m+1}^e) + \tilde{g}_{ii}^e(z_i^e, z_{m+1}^e)v_i^e + \sum_{j=m+1}^{2m} \tilde{g}_{ij}^e(z_i^e, z_{m+1}^e)v_j^e, \qquad i = 1, \ldots, m,\\
\dot{z}_{m+1}^e &= \tilde{f}_{m+1}^e(z_{m+1}^e) + \sum_{j=m+1}^{2m} \tilde{g}_{m+1,j}^e(z_{m+1}^e)v_j^e,\\
y_i &= h_i^e(z_i^e, z_{m+1}^e), \qquad i = 1, \ldots, m,
\end{aligned} \tag{4.53}$$

with $\mathcal{R}_i^e = \operatorname{span}\{\partial/\partial z_i^e\}$. This can be proved as in proposition 4.3, noting that $\mathcal{R}_i^e \cap \sum_{j\neq i} \mathcal{R}_j^e = 0$ for $i = 1, \ldots, m$ and for all x^e in a neighbourhood of x_0^e.

Since $\tilde{f}^e(x_0^e) = 0$, $\tilde{g}_i^e \in \mathcal{R}_i^e$ and $\mathcal{R}_i^e$ is locally invariant under $\tilde{f}^e, \tilde{g}_1^e, \ldots, \tilde{g}_m^e$, it makes sense to consider the restrictions $\tilde{f}^e|\mathcal{L}_{x_0^e}^{\mathcal{R}_i^e}$ and $\tilde{g}_i^e|\mathcal{L}_{x_0^e}^{\mathcal{R}_i^e}$. These vector fields define a control system on $\mathcal{L}_{x_0^e}^{\mathcal{R}_i^e}$ which, in z^e–coordinates, is given by

$$\dot{z}_i^e = \tilde{f}_i^e(z_i^e, 0) + \tilde{g}_{ii}^e(z_i^e, 0)v_i^e. \tag{4.54}$$

We claim that (4.54) is locally asymptotically stabilizable at the origin via dynamic feedback. It will suffice to show that (4.54) is locally diffeomorphic to (4.8). For, let π be the natural projection $\pi : \mathbb{R}^{n+n^w} \to \mathbb{R}^n$. From Chow's theorem ([14] or [55], lemma 2.1.10), x_0^e can be joined to each point of $\mathcal{L}_{x_0^e}^{\mathcal{R}_i^e}$ by a concatenation of integral curves of the vector fields X_{ik}^e, $k = 1, \ldots, s_i$. These integral curves are described by the following set of differential equations

$$\begin{aligned}
\dot{x}_i &= Y_{ik}(x_i),\\
\dot{x}_j &= 0, \qquad j \neq \{i, (m+1, 1)\},\\
\dot{x}_{m+1,1} &= Z_{ik}(0, \ldots, x_i, 0, \ldots, 0, x_{m+1,1}),\\
\dot{\lambda}_i &= Y_{ik}(x_i),\\
\dot{\mu}_i &= Z_{ik}(0, \ldots, x_i, 0, \ldots, 0, \mu_i).
\end{aligned} \tag{4.55}$$

Thus, $x_j = 0$ for $j \notin \{i, (m+1, 1)\}$, since by definition $\mathcal{L}_{x_0^e}^{\mathcal{R}_i^e}$ passes through $x_0^e = 0$. Moreover, from (4.55) $\dot{x}_i = \dot{\lambda}_i$ and, thus, $x_i = \lambda_i$ on $\mathcal{L}_{x_0^e}^{\mathcal{R}_i^e}$, since $x_0^e = 0$. Since $x_j = 0$ for $j \notin \{i, (m+1, 1)\}$ on $\mathcal{L}_{x_0^e}^{\mathcal{R}_i^e}$, we have from (4.55)

$$\dot{x}_{m+1,1} = Z_{ik}(0, \ldots, 0, x_i, 0, \ldots, 0, x_{m+1,1}), \tag{4.56}$$

$$\dot{\mu}_i = Z_{ik}(0, \ldots, 0, x_i, 0, \ldots, 0, \mu_i). \tag{4.57}$$

But, since $x_{m+1,1}$ and μ_i have the same dimension and the initial conditions of the trajectories of (4.56) and (4.57), respectively, are always equal on $\mathcal{L}_{x_0^e}^{\mathcal{R}_i^e}$, it follows that $x_{m+1,1} = \mu_i$ on $\mathcal{L}_{x_0^e}^{\mathcal{R}_i^e}$. This implies that, locally, π maps diffeomorphically $\mathcal{L}_{x_0^e}^{\mathcal{R}_i^e}$ onto $\mathcal{L}_{x_0}^{\mathcal{R}_i^*}$. This, together with the fact that $\pi_{*x^e}\widetilde{f}^e(x^e) = \widetilde{f}(\pi(x^e))$ and $\pi_{*x^e}\widetilde{g}_i^e(x^e) = \widetilde{g}_i(\pi(x^e))$ for all x^e in a neighbourhood of x_0^e (see remark 4.15), proves our claim.

We will show now that the system

$$\dot{z}_{m+1}^e = \widetilde{f}_{m+1}^e(z_{m+1}^e) + \sum_{j=m+1}^{2m} \widetilde{g}_{m+1,j}^e(z_{m+1}^e)v_j^e \tag{4.58}$$

is locally asymptotically stabilizable at the origin via static feedback. It will suffices to show that the matrix

$$\bar{G}(z_{m+1}^e) = (\, \widetilde{g}_{m+1,m+1}^e(z_{m+1}^e) \quad \cdots \quad \widetilde{g}_{m+1,2m}^e(z_{m+1}^e)\,)$$

has full row rank for all z_{m+1}^e in a neighbourhood of $z_{m+1,0}^e = 0$ or, equivalently, that the system (4.58) with output vector z_{m+1}^e has vector relative degree one at $z_{m+1,0}^e$. For, note that the matrix $\bar{G}(z^e)$ has n^w columns and $n + n^w - \sum_{i=1}^{m} s_i \leq n^w$ rows. Denoting by rank $\{A(z^e)\}$ the rank of a given matrix $A(z^e)$, we have rank $\{\bar{G}(z_{m+1}^e)\} = n^w - \dim(\ker\{\bar{G}\}(z^e))$. If $\dim(\ker\{\bar{G}\}(z^e)) > \sum_{i=1}^{m} s_i - n$, it would follow that $\dim(((\sum_{i=1}^{m} \mathcal{R}_i^e) \cap \mathcal{G}^e)(z^e)) > \sum_{i=1}^{m} s_i - n + m$. Since

$$\begin{aligned}
&\dim(((\sum_{i=1}^{m} \mathcal{R}_i^e) \cap \mathcal{G}^e)(z^e)) = \dim(((\sum_{i=1}^{m} \mathcal{R}_i^e) \cap \mathcal{G}^w)(z^e)) + \\
&+ \dim(((\sum_{i=1}^{m} \mathcal{R}_i^e) \cap (\mathrm{span}\,\{\widetilde{g}_1^e, \ldots, \widetilde{g}_m^e\}))(z^e)) = \dim(((\sum_{i=1}^{m} \mathcal{R}_i^e) \cap \mathcal{G}^w)(z^e)) + \\
&+ \dim((\mathrm{span}\,\{\widetilde{g}_1^e, \ldots, \widetilde{g}_m^e\})(z^e)) = \dim(((\sum_{i=1}^{m} \mathcal{R}_i^e) \cap \mathcal{G}^w)(z^e)) + m,
\end{aligned}$$

we obtain $\dim(((\sum_{i=1}^{m} \mathcal{R}_i^e) \cap \mathcal{G}^w)(z^e)) > \sum_{i=1}^{m} s_i - n$. We claim that $\dim(((\sum_{i=1}^{m} \mathcal{R}_i^e) \cap \mathcal{G}^w)(z^e)) = \sum_{i=1}^{m} s_i - n$, which clearly gives a contradiction. Indeed, denoting by $X(x)$ the matrix

$$(\, X_{11}(x) \quad \cdots \quad X_{1s_1}(x) \quad \cdots \quad X_{m1}(x) \quad \cdots \quad X_{ms_m}(x)\,)$$

and by $X^e(z^e)$ the matrix

$$(\, X_{11}^e(z^e) \quad \cdots \quad X_{1s_1}^e(z^e) \quad \cdots \quad X_{m1}^e(z^e) \quad \cdots \quad X_{ms_m}^e(z^e)\,),$$

since $\dim((\mathrm{span}\{X\})(x)) = n$ and $\dim((\mathrm{span}\{X^e\})(z^e)) = \sum_{i=1}^{m} s_i$, it follows that dim $((\mathrm{span}\{X^e\} \cap \mathcal{G}_w)(z^e))$ and $\dim((\ker\{X\})(x))$ are constant in a neighbourhood of $z^e(x_0^e)$ and x_0, respectively. Moreover, $\dim((\mathrm{span}\{X^e\} \cap \mathcal{G}^w)(z^e)) = \dim((\ker\{X\})(x)) = \sum_{i=1}^{m} s_i - n$, which proves our claim.

If we choose

$$\begin{aligned} v_i^e &= \varphi_i^e(z_i^e, \eta_i^e), \\ \dot{\eta}_i^e &= \psi_i^e(z_i^e, \eta_i^e), \end{aligned}$$

such that the system, resulting from (4.54), is locally asymptotically stable at the origin and $v_i^e = \varphi_i(z_{m+1}^e)$, $i = m+1, \ldots, 2m$, such that the system, resulting from (4.58), is locally asymptotically stable at the origin, then also the closed-loop system

$$\begin{aligned} \dot{z}_i^e &= \widetilde{f}_i^e(z_i^e, z_{m+1}^e) + \widetilde{g}_{ii}^e(z_i^e, z_{m+1}^e)\varphi_i^e(z_i^e, \eta_i^e) + \sum_{j=m+1}^{2m} \widetilde{g}_{ij}^e(z_i^e, z_{m+1}^e)\varphi_j^e(z_{m+1}^e),\ i = 1, \ldots, m, \\ \dot{\eta}_i^e &= \psi_i^e(z_i^e, \eta_i^e), \qquad i = 1, \ldots, m, \\ \dot{z}_{m+1}^e &= \widetilde{f}_{m+1}^e(z_{m+1}^e) + \sum_{j=m+1}^{2m} \widetilde{g}_{m+1,j}^e(z_{m+1}^e)\varphi_j^e(z_{m+1}^e) \end{aligned}$$

is locally asymptotically stable at the origin. Moreover, as it can be easily checked, the system, obtained from plugging the feedback law

$$\begin{aligned} v_i^e &= \varphi_i^e(z_i^e, \eta_i^e) + v_i, & i &= 1, \ldots, m, \\ \dot{\eta}_i^e &= \psi_i^e(z_i^e, \eta_i^e), & i &= 1, \ldots, m, \\ v_i^e &= \varphi_i^e(z_{m+1}^e), & i &= m+1, \ldots, 2m, \end{aligned}$$

into (4.53), is noninteractive and has vector relative degree at the origin of the state space. □

As it results from the the proof of theorem 4.16, the feedback law, obtained by cascading $u^e = \alpha^e(x^e) + \sum_{j=i}^{m} \beta_j^e(x^e)v_j^e$ (proposition 4.12) together with (4.52), solves NLSD. This feedback law is invertible, since, as it can be easily checked, it has the form

$$\begin{aligned} u_i &= \alpha_i(x, w, \eta_i^e) + v_i, \qquad i = 1, \ldots, m, \\ \dot{w} &= \delta(x, w, \eta_i^e) + \sum_{j=1}^{m} \gamma_j(x, w, \eta_i^e)v_j, \\ \dot{\eta}_i^e &= \psi_i^e(x, w, \eta_i^e), \qquad i = 1, \ldots, m. \end{aligned}$$

We sum up all the above results in the following theorem.

Theorem 4.17. Assume that (A1), (A2), (A3) and (A4) hold. If

a) the dynamics (4.7) is locally asymptotically stable at the origin,

b) each system (4.8) is locally asymptotically stabilizable at the origin via dynamic feedback,

then NLSD is solvable. On the other hand, let (1.2) be a feedback law which solves NLSD. If, in addition, the distributions $\mathcal{R}_i^e = \langle \tilde{f}^e, \tilde{g}_1^e, \ldots, \tilde{g}_m^e | \mathrm{span}\{\tilde{g}_i^e\}\rangle$, $i = 1, \ldots, m$, *are regularly computable at* x_0^e *and the distributions* $\mathcal{R}_i^e + \Delta_{MIX}^e$, $i = 1, \ldots, \nu$, *and* Δ_{MIX}^e *have constant dimension in a neighbourhood of* x_0^e, *then a) and b) are satisfied.* □

Remark 4.18. Under assumptions (A1), (A2), (A3) and (A4), conditions a) and b) of theorem 3.5 imply a) and b) of theorem 4.18. Indeed, if the dynamics (3.5) is locally asymptotically stable at the origin, since (4.7) is an invariant subdynamics of (3.5), it must be locally asymptotically stable at the origin. On the other hand, by assumption each system (3.6) is locally asymptotically stabilizable at the origin via static feedback or, equivalently, with the local decomposition (4.6), each system

$$\dot{z}_1^i = f_{z_1^i}(0, \ldots, 0, z_1^i) + g_{z_1^i}(0, \ldots, 0, z_1^i)u_i$$

is locally stabilizable at the origin via static feedback law. Moreover, if (3.5) is locally asymptotically stable, from (4.6) it follows that also the invariant subdynamics of (3.5)

$$\dot{z}_3^i = f_{z_3^i}(0, \ldots, 0, z_3^i)$$

is locally asymptotically stable at the origin. This implies that the system

$$\begin{aligned} \dot{z}_1^i &= f_{z_1^i}(0, \ldots, 0, z_1^i) + g_{z_1^i}(0, \ldots, z_1^i)u_i, \\ \dot{z}_3^i &= f_{z_3^i}(0, \ldots, 0, z_1^i, 0, z_3^i) + g_{z_3^i}(0, \ldots, 0, z_1^i, 0, z_3^i)u_i, \end{aligned}$$

which is exactly (4.8), is locally asymptotically stabilizable at the origin via static feedback (in particular, via dynamic feedback). □

Remark 4.19. In the case of linear systems, $\Delta_{MIX}^* = 0$ and, thus, condition a) of theorem 4.17 is trivially satisfied. Also condition b) is satisfied by definition of controllability subspace. Since (A, B) is controllable, it follows from proposition 3.1 that $\sum_{j=1}^{m} \mathcal{R}_j^* = \mathbb{R}^n$. This, together with $\sum_{j \neq i} \mathcal{R}_j^* \subset \mathcal{K}_i$, implies that $\mathcal{R}_i^* + \mathcal{K}_i = \mathbb{R}^n$ for $i = 1, \ldots, m$, which is the necessary and sufficient condition of theorem **2.3.4**. □

3.4.3 Efficient noninteracting control with stability

In section 2.4.2 we have proposed a solution of NLSD without imposing any constraint on the dimension n^w of the dynamic extension. In that case, n^w has been chosen equal to $m(n_i + n_0) = n + n_0(m-1)$. In this section, we will consider some cases in which n^w can be reduced. A first attempt is

$$n^w = \sum_{i=1}^{m} s_i$$

which is less or equal than $m(n_i+n_0)$. Note that $\mathrm{span}\{\partial/\partial x_i\} \subset \mathcal{R}_i^* + \mathrm{span}\{\partial/\partial x_{m+1,1}\}$ in a neighbourhood of x_0 (this follows from $\sum_{i=1}^{m} \mathcal{R}_i^* = I\!R^n$ and $\mathcal{R}_i^* = \langle \widetilde{f}, \widetilde{g}_1, \ldots, \widetilde{g}_m | \mathrm{span}\{\widetilde{g}_i\}\rangle$) and that the vector fields X_{ik}, $k = 1, \ldots, s_i$, are of the form (4.26). As a consequence, from the $(n_i + n_0) \times s_i$ matrix

$$X_i(x) = \begin{pmatrix} Y_{i1}(x_i) & \ldots & Y_{is_i}(x_i) \\ Z_{i1}(x) & \ldots & Z_{is_i}(x) \end{pmatrix}$$

it is always possible to choose a $s_i \times s_i$ submatrix of the form

$$\bar{X}_i(x) = \begin{pmatrix} Y_{i1}(x_i) & \ldots & Y_{is_i}(x_i) \\ \bar{Z}_{i1}(x) & \ldots & \bar{Z}_{is_i}(x) \end{pmatrix},$$

which is invertible for all x in a neighbourhood of x_0.

For simplicity, we will assume that $\bar{X}_i(x)$ is given by the first s_i rows of $X_i(x)$ and denote by $\bar{X}_{ik}(x)$ the k-th column of $\bar{X}_i(x)$. Let $x^i_{m+1,1}$ be the first $s_i - n_i$ components of $x_{m+1,1}$ and $\bar{x}^i_{m+1,1}$ be the last $n_0 - (s_i - n_i)$ ones. Moreover, let $\dim w_i = s_i$, $\dim \mu_i = s_i - n_i$, $i = 1, \ldots, m$. Define the following vector fields $X^e_{ik} \in I\!R^{n+n^w}$

$$X^e_{ik}(x^e) = \begin{pmatrix} X_{ik}(x) \\ 0_{s_1 \times 1} \\ \vdots \\ 0_{s_{i-1} \times 1} \\ X^*_{ik}(x_i, \mu_i) \\ 0_{s_{i+1} \times 1} \\ \vdots \\ 0_{s_m \times 1} \end{pmatrix}, \qquad i = 1, \ldots, m,\ k = 1, \ldots, s_i,$$

where

$$X^*_{ik}(x_i, \mu_i) = \bar{X}_{ik}(x)\big|_{x_j = 0 \text{ for } j \notin \{i,(m+1,1)\}\ ;\ x^i_{m+1,1} = \mu_i} \tag{4.59}$$

and, correspondingly, define the distributions

$$\mathcal{R}^e_i = \mathrm{span}\{X^e_{ik} : k = 1, \ldots, s_i\}, \qquad i = 1, \ldots, m.$$

The distributions $\mathcal{R}_i^e$, $i = 1, \ldots, m$, defined above, are clearly independent in a neighbourhood of x_0^e, since otherwise the vector fields X_{ik}, $k = 1, \ldots, s_i$, would not be linearly independent in a neighbourhood of x_0. Moreover, $\mathcal{R}_i^e$ is also locally involutive if we assume that

(R1) $\frac{\partial X_{ik}}{\partial \bar{x}^i_{m+1,1}}(x) = 0$ *for* $i = 1, \ldots, m$ *and for all* x *in a neighbourhood of* x_0.

As a matter of fact, if assumption (R1) holds, we have

$$[X_{ik}^*, X_{ih}^*](x_i, \mu_i) = [X_{ik}, X_{ih}]^*(x_i, \mu_i),$$

where $[X_{ik}, X_{ih}]^*(x_i, \mu_i)$ is defined as in (4.59) but with $X_{ik}(x)$ replaced by $[X_{ik}, X_{ih}](x)$. From here, we can proceed as in proposition 4.7.

If $p = 2$, assumption (R1) is trivially satisfied, since in this case $\mathcal{R}_i^* = \text{span}\{\partial/\partial x_i, \partial/\partial x_{m+1,1}\}$.

Next, we will try to further reduce n^w by picking

$$n^w = \sum_{i=2}^{m} \dim(\mathcal{R}_i^* \cap \sum_{j<i} \mathcal{R}_j^*),$$

which is clearly less than $\sum_{i=1}^m s_i$, since $\mathcal{R}_i^* \cap \sum_{j<i} \mathcal{R}_j^* \subset \mathcal{R}_i^*$. We will assume that

(R2) the distributions $\mathcal{R}_i^* \cap (\sum_{j<i} \mathcal{R}_j^*)$, $i = 2, \ldots, m$, *have constant dimension* ϱ_i *for all* x *in a neighbourhood of* x_0 *and there exist vector fields* $X_{i1}, \ldots, X_{i\varrho_i}$ *of the form (4.23) such that* $\mathcal{R}_i^* \cap (\sum_{j<i} \mathcal{R}_j^*) = \text{span}\{X_{i1}, \ldots, X_{i\varrho_i}\}$ *for* $i = 2, \ldots, m$ *and for all* x *in a neighbourhood of* x_0.

Since $\varrho_i < s_i$, it follows that there exist vector fields $X_{i1}, \ldots, X_{is_i}$ of the form (4.23) such that $\mathcal{R}_i^* = \text{span}\{X_{i1}, \ldots, X_{is_i}\}$ and$\mathcal{R}_i^* \cap (\sum_{j<i} \mathcal{R}_j^*) = \text{span}\{X_{i1}, \ldots, X_{i\varrho_i}\}$ in a neighbourhood of x_0.

Since the vector fields X_{ik}, $k = 1, \ldots, s_i$ have the form (4.26), it is always possible to choose from the $n_0 \times s_i$ matrix

$$Z_i(x) = (\, Z_{i1}(x) \quad \ldots \quad Z_{is_i}(x)\,)$$

ϱ_i rows $w_{i_1}(x), \ldots, w_{i_{\varrho_i}}(x)$, which are linearly independent at each x in a neighbourhood of x_0 and such that, after possibly renumbering the columns of $Z_i(x)$, the matrix

$$\bar{Z}_i(x) = \begin{pmatrix} w_{i_1}(x) \\ \vdots \\ w_{i_{\varrho_i}}(x) \end{pmatrix}$$

has the first ϱ_i columns linearly independent at each x in a neighbourhood of x_0. For simplicity, we assume that the rows of $\bar{Z}_i(x)$ are given by the first ϱ_i rows of $Z_i(x)$ and denote by $\bar{Z}_{ik}(x)$ the k–th column of $\bar{Z}_i(x)$. Let $x^i_{m+1,1}$ be the first ϱ_i components of $x_{m+1,1}$ and $\bar{x}^i_{m+1,1}$ be the last $n_0 - \varrho_i$ ones. Moreover, choose $\dim w_1 = 0$, $\dim w_i = \dim \mu_i = \varrho_i$, $i = 2, \dots, m$. Let us define the following vector fields $X^e_{ik} \in I\!R^{n+n^w}$

$$X^e_{1k}(x^e) = \begin{pmatrix} X_{1k}(x) \\ 0_{n^w \times 1} \end{pmatrix}, \qquad X^e_{ik}(x^e) = \begin{pmatrix} X_{ik}(x) \\ 0_{\varrho_1 \times 1} \\ \vdots \\ 0_{\varrho_{i-1} \times 1} \\ X^*_{ik}(x_i, \mu_i) \\ 0_{\varrho_{i+1} \times 1} \\ \vdots \\ 0_{\varrho_m \times 1} \end{pmatrix} \qquad i = 2, \dots, m, \; k = 1, \dots s_i,$$

where

$$X^*_{ik}(x_i, \mu_i) = \bar{Z}_{ik}(x)\big|_{x_j = 0 \text{ for } j \notin \{i, m+1\}\ ;\ x^i_{m+1,1} = \mu_i} \qquad ,$$

and, correspondingly, define the distributions

$$\mathcal{R}^e_i = \operatorname{span}\{X^e_{ik} : k = 1, \dots, s_i\}, \qquad i = 1, \dots, m.$$

Each distribution $\mathcal{R}^*_i$ is easily shown to be involutive, if we assume that

(R3) $\frac{\partial \bar{Z}_{ik}}{\partial \bar{x}^i_{m+1,1}}(x) = 0$ *for* $i = 2, \dots, m$ *and for all* x *in a neighbourhood of* x_0.

On the other hand, to prove that $\mathcal{R}^e_1, \dots, \mathcal{R}^e_m$ are linearly independent in a neighbourhood of x_0,it suffices to prove that the matrix

$$X^e(x^e) = (\, X^e_{11}(x^e) \quad \cdots \quad X^e_{1s_1}(x^e) \quad \cdots \quad X^e_{m1}(x^e) \quad \cdots \quad X^e_{ms_m}(x^e)\,)$$

has rank equal to $\sum_{i=1}^m s_i$ for all x^e in a neighbourhood of x^e_0. If this were not true, there would exist real numbers c_{jk} such that

$$\sum_{j=1}^{m} \sum_{k=1}^{s_j} c_{jk} X_{jk}(0) = 0, \tag{4.60}$$

$$\sum_{k=1}^{s_i} c_{jk} X^*_{jk}(0,0) = 0, \qquad j = 2, \dots, m, \tag{4.61}$$

with at least one c_{jk} is nonzero. Let i be the greatest integer such that $c_{ik} \neq 0$ for some k. From (4.60)

$$\sum_{j<i} \sum_{k=1}^{s_j} c_{jk} X_{jk}(0) = \sum_{k=1}^{s_i} c_{ik} X_{ik}(0),$$

which implies that there exists a nonzero vector $v \in \mathcal{R}_i^*(x_0) \cap \sum_{j<i} \mathcal{R}_j^*(x_0) = (\mathcal{R}_i^* \cap \sum_{j<i} \mathcal{R}_j^*)(x_0)$ (by $\Delta(x_0)$ we denote the subspace assigned by the distribution Δ at x_0). From assumption (R2) it follows that $c_{ik} = 0$ for $k > \varrho_i$. But this, together with (4.61), since the first ϱ_i columns are linearly independent by construction, implies that $c_{ik} = 0$ for all k, which is clearly a contradiction.

It is worth noting that, if we want to define the distributions $\mathcal{R}_1^e, \ldots, \mathcal{R}_m^e$ in such a way that they are linearly independent in a neighbourhood of x_0^e, we *cannot* take n^w less than $\sum_{i=2}^{m} \dim(\mathcal{R}_i^* \cap \sum_{j<i} \mathcal{R}_j^*)$. As a matter of fact, it must be

$$n + n^w - \sum_{i=1}^{m} s_i \geq 0,$$

which implies

$$\begin{aligned} n^w \geq \sum_{i=1}^{m} s_i - n &= \sum_{i=1}^{m} \dim \mathcal{R}_i^* - \dim(\sum_{i=1}^{m} \mathcal{R}_i^*) = \\ &= \sum_{i=2}^{m} \dim(\mathcal{R}_i^* \cap (\sum_{j<i} \mathcal{R}_j^*)). \end{aligned}$$

The last equality can be proved exactly as in [91] (lemma 9.3). If $p = 2$, we have $\varrho_2 = n_0$ and assumption (R3) is trivially satisfied.

3.4.4 Necessary and sufficient conditions

In this section, we will turn again our attention to non–efficient solutions of NLSD. We will see that, given *any* dynamic feedback law which solves ND for (1.1), the system, resulting from (1.1) after feedback, contains a subsystem, which is obtained from (1.1) itself by means of a *canonical* dynamic extension and has vector relative degree at the origin of the state space. This allows to *uniquely* (in some sense) associate a dynamically extended system of the form (3.8) to a given system (1.1), for which ND is solvable, and to define for this system the corresponding dynamics (4.7) and (4.8) (in a unique way). At this point, theorem 4.17 does the rest.

Necessary and sufficient conditions to solve ND are well–known and can be characterized in terms of a suitable algorithm ([55], section 7.7.5), which will be shortly reviewed for reader's convenience (the notation is slightly different from the one used in [55]). Let Σ^0 be (1.1) and let $x^0 = x, x_0^0 = x_0, f^0(x^0) = f(x), G^0(x^0) = G(x), h_i^0(x^0) = h_i(x)$, $i = 1, \ldots, m$, and $v^0 = u$. Moreover, let $U^0 = \mathbb{R}^n$.

Dynamic extension algorithm. (step $k \geq 0$). Let us consider the decoupling matrix $A^k(x^k)$ of Σ^k. Assume that the rank of $A^k(x^k)$ is constant on an open and dense subset $U^{k+1} \subset U^k$. Let q^k be this rank with $q^k < m$ (otherwise, Σ^k has already vector relative degree $r_1, \ldots, r_m$ for all $x^k \in U^{k+1}$). After possibly renumbering the outputs, we can suppose that the first q^k rows of $A^k(x^k)$ are linearly independent for all $x^k \in U^{k+1}$. Let $v^k = \alpha^k(x^k) + \beta^k(x^k)v^{k+1}$ be a regular feedback law such that, denoting by β_j^k the j–th column of β^k, we have

$$L_{f^k+G^k\alpha^k} L_{f^k}^{r_i-1} h_i^k(x^k) = 0 \qquad 1 \leq i \leq q^k,$$

$$L_{G^k\beta_j^k} L_{f^k}^{r_i-1} h_i^k(x^k) = \delta_{ij}, \qquad 1 \leq i \leq q^k \,,\; 1 \leq j \leq m,$$

for all $x^k \in U^{k+1}$ (δ_{ij} is the Kronecker's delta). Let

$$\zeta^k = \begin{pmatrix} \zeta_1^k \\ \vdots \\ \zeta_{q^k}^k \end{pmatrix} \,,\; x^{k+1} = \begin{pmatrix} x^k \\ \zeta^k \end{pmatrix} \,,\; h_i^{k+1}(x^{k+1}) = h_i^k(x^k) \,,$$

$$(v')^{k+1} = \begin{pmatrix} v_1^{k+1} \\ \vdots \\ v_{q^k}^{k+1} \end{pmatrix} \,,\; (v'')^{k+1} = \begin{pmatrix} v_{q^k+1}^{k+1} \\ \vdots \\ v_m^{k+1} \end{pmatrix} \,,$$

and denote by $(\beta')^k(x^k)$ (respectively, by $(\beta'')^k(x^k)$) the matrix given by the first q^k columns (respectively, by the last $m - q^k$ columns) of $\beta^k(x^k)$. Now, define the following feedback law

$$v^k = \alpha^k(x^k) + (\beta')^k(x^k)\zeta^k + (\beta'')^k(x^k)(v'')^{k+1},$$

$$\dot{\zeta}^k = (v')^{(k+1)},$$

and denote by Σ^{k+1} the system

$$\dot{x}^k = f^k(x^k) + G^k(x^k)\alpha^k(x^k) + G^k(x^k)(\beta')^k(x^k)\zeta^k +$$

$$+ G^k(x^k)(\beta'')^k(x^k)(v'')^{k+1},$$

$$\dot{\zeta}^k = (v')^{k+1},$$

$$y_i = h_i^{k+1}(x^{k+1}), \qquad i = 1, \ldots, m.$$

We want to remark that the dynamic extension algorithm does not produce, in general, a dynamic feedback law (or dynamic *compensator*) with *minimal* dimension. On the other hand, an algorithm with such property is the celebrated Descusse and Moog's algorithm ([26]).

The following propositions completely characterize the relationships between the dynamic extension algorithm and the problem of achieving vector relative degree via dynamic feedback laws. The proof is found in [55] (section 7.7.5).

Proposition 4.20. Assume that $A^h(x^h)$, $h \geq 0$, has constant rank on an open and dense subset of its domain. For each fixed $k \geq 0$, only one of the following facts can happen:

(i) if $l = n - (r_1 + \ldots + r_{q^k} + \min\{r_j : q^k + 1 \leq j \leq m\})$, the matrix $A^{k+l+1}(x^{k+l+1})$ has rank greater or equal to $q^k + 1$ for all x^{k+l+1} in an open and dense subset of its domain;

(ii) for all $h \geq 0$ the matrix $A^{k+h}(x^{k+h})$ has rank equal to q^k for all x^{k+h} in an open and dense subset of its domain. □

Proposition 4.21. Assume that $A^h(x^h)$, $h \geq 0$, has constant rank on an open and dense subset of its domain. If there exist $n^w \geq 0$ and a dynamic feedback law such that the resulting closed-loop system

$$
\begin{aligned}
\dot{x} &= f(x) + G(x)\alpha(x,w) + G(x)\beta(x,w)v, \\
\dot{w} &= \gamma(x,w) + \delta(x,w)v, \qquad w \in I\!\!R^{n^w}, \\
y_i &= h_i(x), \qquad i = 1,\ldots,m,
\end{aligned}
\tag{4.62}
$$

has vector relative degree at each point (x°, w°) of an open and dense subset of $I\!\!R^{n+n^w}$, then, after a finite number $k^* \geq 0$ of iterations, the dynamic extension algorithm produces a system Σ^{k^*} with a decoupling matrix $A^{k^*}(x^{k^*})$ nonsingular at each point of an open and dense subset of its domain. □

In other words, the dynamic extension algorithm converges with $q_{k^*} = m$ for some $k^* \geq 0$ and for all x^{k^*} in an open and dense subset of the domain of definition of $A^{k^*}(x^{k^*})$ if and only if there exist $n^w \geq 0$ and a dynamic feedback law such that (4.62) has vector relative degree at each point (x, w) of an open and dense subset of $I\!\!R^{n+n^w}$. If $f(x), g_j(x)$ and $h_j(x)$, $j = 1, \ldots, m$ are analytic functions of x, $A^h(x^h)$, $h \geq 0$, has *always* constant rank on some open and dense subset of its domain. In what follows, we will assume that $A^h(x^h)$, $h \geq 0$, has constant rank in an open neighbourhood of x_0^h and that there exists $k^* \geq 0$ such that $q_{k^*} = m$. In other words, we will consider in a natural way the class of nonlinear systems for which (modulo singularities) ND is solvable.

Now, we define a *canonical* dynamic extension of (1.1) by means of the following algorithm ([92]). Similarly to Descusse and Moog's algorithm ([26]), this algorithm produces a *minimal* compensator such that the resulting closed-loop system (4.62) has relative degree at x_0^e. Let Σ_C^0 be (1.1) and $x_C^0 = x, x_{C0}^0 = x_0, f_C^0(x_C^0) = f(x), g_{Ci}^0(x_C^0) = g_i(x)$, $u_{Ci}^0 = u_i$ and $h_{Ci}^0(x_C^0) = h_i(x)$ for $i = 1, \ldots, m$.

Canonical dynamic extension algorithm. (Step $k \geq 0$) Consider the system Σ_C^k

$$\dot{x}_C^k = f_C^k(x_C^k) + \sum_{j=1}^{m} g_{Cj}^k(x_C^k)u_{Cj}^k,$$
$$y_i = h_{Ci}^k(x_C^k), \qquad i = 1, \ldots, m.$$

Assume that the decoupling matrix $A_C^k(x_C^k)$ of Σ_C^k has constant rank for all x^k in a neighbourhood U_C^k of x_{C0}^k. If $A_C^k(x_C^k)$ has rank m for all x^k in U_C^k, then Σ_C^k has vector relative degree at x_{C0}^k. Otherwise, after possibly renumbering the outputs, denoting by $A_j^k(x_C^k)$ the j–th row of $A_C^k(x_C^k)$, we have

$$\text{rank}\left\{\begin{pmatrix} A_1^k(x_C^k) \\ \vdots \\ A_{j_k-1}^k(x_C^k) \end{pmatrix}\right\} = \text{rank}\left\{\begin{pmatrix} A_1^k(x_C^k) \\ \vdots \\ A_{j_k}^k(x_C^k) \end{pmatrix}\right\} = j_k - 1$$

for some $1 < j_k \leq m$ and for all $x_C^k \in U_C^k$. Thus, if $a_{ij}^k(x_C^k)$ denotes the i–th component of $A_j^k(x_C^k)$, there exist smooth real–valued functions $c_i^k(x_C^k)$, two intergers $i_k \leq j_k - 1$ and l_k such that

$$A_{j_k}^k(x_C^k) = \sum_{i=1}^{j_k-1} c_i^k(x_C^k)A_i^k(x_C^k)$$

with $c_{i_k}^k(x_{C0}^k) \neq 0$ and $a_{i_k l_k}^k(x_{C0}^k) \neq 0$. Define the following feedback law

$$\begin{aligned} \dot{\xi}^k &= u_{Cl_k}^{k+1}, \\ u_{Cl_k}^k &= \frac{1}{a_{i_k l_k}^k(x_C^k)}[p^k(x_C^k)(\xi^k - q^k(x_C^k)) - \sum_{j \neq l_k} a_{i_k j}^k(x_C^k)u_{Cj}^k], \end{aligned} \tag{4.63}$$

where $p^k(x_C^k)$ and $q^k(x_C^k)$ are arbitrary smooth real–valued functions such that $p^k(x_{C0}^k) \neq 0$ and $q^k(x_{C0}^k) = 0$. Let Σ_C^{k+1}:

$$\dot{x}_C^{k+1} = f_C^{k+1}(x_C^{k+1}) + \sum_{j=1}^{m} g_{Cj}^{k+1}(x_C^{k+1})u_{Cj}^{k+1},$$
$$y_i = h_{Ci}^{k+1}(x_C^{k+1}), \qquad i = 1, \ldots, m,$$

be the system obtained from applying (4.63) to Σ_C^k with

$$x_C^{k+1} = ((x_C^k)^T \quad (\xi_C^k)^T)\, ,\; \xi_{C0}^k = 0\, ,\; x_{C0}^{k+1} = ((x_{C0}^k)^T \quad (\xi_{C0}^k)^T)$$

and

$$u_{Ci}^{k+1} = u_{Ci}^k \text{ for } i \neq l_k.$$

Note that, if $f(x), g_i(x)$ and $h_i(x)$, $i = 1, \ldots, m$ are analytic functions of x, then the matrices $A_C^k(x_C^k)$, $k \geq 0$, have *always* constant rank on an open and dense set, not necessarily containing x_{C0}^k. Let us consider a dynamic feedback law which solves ND for (1.1) and denote by $\widetilde{\Sigma}^e$ the corresponding closed-loop system (4.62). Moreover, assume that $A_C^k(x_C^k)$ has constant rank for all k and for all x_C^k in a neighbourhood of x_{C0}^k. The system $\widetilde{\Sigma}^e$ can be considered as the cascade of the system $\Sigma_C^{k_C^*}$, obtained from the canonical dynamic extension algorithm after a finite number $k_C^* \geq 0$ of steps and having itself vector relative degree at each $x_C^{k_C^*}$ in a neighbourhood of $x_{C0}^{k_C^*}$, together with an invertible system Σ_F (see Figure 4.2).

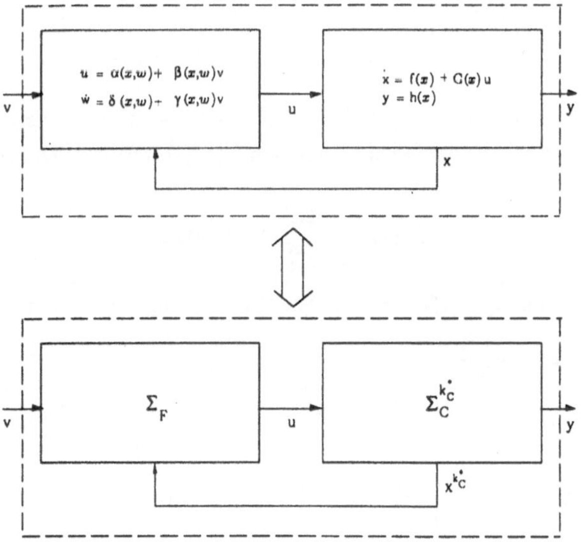

Figure 4.2

In other words, $\Sigma_C^{k_C^*}$ is a dynamic extension of (1.1) with *minimum* dimension and every other system, which has been obtained from (1.1) through invertible dynamic feedback laws and has vector relative degree at the origin of the state space (in particular, the

system Σ^{k^*} obtained after applying the dynamic extension algorithm), is *equal* to $\Sigma_C^{k_C^*}$ up to local changes of coordinates and invertible feedback tranformations. Indeed, reasoning as in [92] (theorem 3.1), it can be shown that, possibly after a local change of coordinates, $\widetilde{\Sigma}^e$ is of the form

$$\begin{aligned}
\dot{x}^{k_C^*} &= f^{k_C^*}(x^{k_C^*}) + \sum_{j=1}^{m} g_j^{k_C^*}(x^{k_C^*})u_j^{k_C^*}, \\
u_i^{k_C^*} &= \alpha_i^{k_C^*}(x^{k_C^*}, z^{k_C^*}) + \sum_{j=1}^{m} \beta_{ij}^{k_C^*}(x^{k_C^*}, z^{k_C^*})v_j, \qquad i = 1, \ldots, m, \\
\dot{z}^{k_C^*} &= \delta^{k_C^*}(x^{k_C^*}, z^{k_C^*}) + \sum_{j=1}^{m} \gamma_j^{k_C^*}(x^{k_C^*}, z^{k_C^*})v_j, \\
y_i &= h_i^{k_C^*}(x^{k_C^*}), \qquad i = 1, \ldots, m,
\end{aligned} \tag{4.64}$$

where

$$\begin{aligned}
\dot{x}^{k_C^*} &= f^{k_C^*}(x^{k_C^*}) + \sum_{j=1}^{m} g_j^{k_C^*}(x^{k_C^*})u_j^{k_C^*}, \\
y_i &= h_i^{k_C^*}(x^{k_C^*}), \qquad i = 1, \ldots, m,
\end{aligned}$$

denotes $\Sigma_C^{k_C^*}$ and

$$\begin{aligned}
u_i^{k_C^*} &= \alpha_i^{k_C^*}(x^{k_C^*}, z^{k_C^*}) + \sum_{j=1}^{m} \beta_{ij}^{k_C^*}(x^{k_C^*}, z^{k_C^*})v_j, \qquad i = 1, \ldots, m, \\
\dot{z}^{k_C^*} &= \delta^{k_C^*}(x^{k_C^*}, z^{k_C^*}) + \sum_{j=1}^{m} \gamma_j^{k_C^*}(x^{k_C^*}, z^{k_C^*})v_j,
\end{aligned} \tag{4.65}$$

is a dynamic feedback law such that $\widetilde{\Sigma}^e$, seen as the composition of $\Sigma_C^{k_C^*}$ together with (4.65), has vector relative degree at x_0^e. Clearly, without loss of generality we can assume that $\Sigma_C^{k_C^*}$ is noninteractive, which, together with proposition 1.5, implies that (4.65) is an invertible feedback law. This proves that $\widetilde{\Sigma}^e$ can be considered as the cascade of $\Sigma_C^{k_C^*}$ together with an invertible system Σ_F. Moreover, since $\Sigma_C^{k_C^*}$ has itself vector relative degree at x_0^e, we can *uniquely* associate to the system (1.1) the dynamics (4.7) and (4.8), correspondingly defined for $\Sigma_C^{k_C^*}$. This assignment is *unique* up to local changes of coordinates and invertible feedback transformations. This motivates the following important result.

Theorem 4.22. Assume that the matrices $A^k(x^k)$ and $A_C^k(x_C^k)$, $k \geq 0$, have constant rank in a neighbourhood of x_0^k and x_{C0}^k, respectively. Then, NLSD is solvable for (1.1) if and only if ND is solvable for (1.1) and NLSD is solvable for $\Sigma_C^{k_C^*}$, the system with vector *relative degree at $x_{C0}^{k_C^*}$, obtained from the canonical dynamic extension algorithm after a finite number $k_C^* \geq 0$ of steps.* □

In the case of linear systems, since the conditions of theorem 4.17 are always satisfied for $\Sigma_C^{k_C^*}$ (remark 4.19), it follows from theorem 4.22 that NSD is solvable if and only if ND is. This is exactly the result stated by theorem 2.3.4.

Note that, as a consequence of the above facts, if (l_k', i_k') and (l_k'', i_k'') are two different choices of the integers l_k and i_k at the k-th step of the canonical dynamic extension algorithm, assuming that in both cases the algorithm can be carried through, we obtain two different systems $\Sigma_C^{(k_C^*)'}$ and $\Sigma_C^{(k_C^*)''}$, which are *equal* up to local changes of coordinates and invertible static feedback transformations, since $\Sigma_C^{(k_C^*)'}$ and $\Sigma_C^{(k_C^*)''}$ must contain each other by minimality.

3.5 Some examples

In this section we will give three examples. The first one is a system of the form (1.1), noninteractive and with vector relative degree at x_0. This system is such that the dynamics (3.5) is unstable or, equivalently, noninteraction and stability *cannot* be achieved via invertible *static* feedback (theorem 3.5). Nonetheless, the dynamics (4.7) is locally asymptotically stable at the origin and each system (4.8) is locally asymptotically stabilizable at the origin via static feedback. Thus, noninteraction and stability *can* be achieved via invertible *dynamic* feedback (theorem 4.17). As a second example, we will see a system for which *neither* NLSS *nor* NLSD are solvable. Finally, we will give the example of a system which does not have vector relative degree at x_0, but satisfies the conditions of theorem 4.22.

Let

$$\begin{aligned}
\dot{x}_1 &= u_1 \\
\dot{x}_2 &= u_2 \\
\dot{x}_3 &= a_1(x_1) + a_2(x_2) + bx_3 \\
\dot{x}_4 &= c(x_1, x_2, x_3, x_4) \\
y_i &= x_i, \qquad i = 1, 2,
\end{aligned} \tag{5.1}$$

be a noninteractive system with $b > 0, a_i(0) = 0, c(0,0,0,0) = 0$, $\frac{\partial a_i}{\partial x_i}(0) \neq 0$ for $i = 1, 2$, $\frac{\partial^{(2)} c}{\partial x_1 \partial x_2}(0,0,0,0) \neq 0$ and $\dot{x}_4 = c(0,0,0,x_4)$ is locally asymptotically stable at $x_4 = 0$ (see

[78]). By straightforward computations we obtain in a neighbourhood of x_0

$$\begin{aligned}\Delta_{MIX} &= \text{span}\{\partial/\partial x_4\},\\ \mathcal{R}^* &= \text{span}\{\partial/\partial x_3, \partial/\partial x_4\},\\ \mathcal{R}_i^* &= \text{span}\{\partial/\partial x_i, \partial/\partial x_3, \partial/\partial x_4\} = \text{span}\{\tilde{g}_i, ad_{\tilde{f}}\tilde{g}_i\} + \Delta_{MIX} \qquad i = 1,2.\end{aligned}$$

The dynamics (3.5), associated to (5.1),

$$\begin{aligned}\dot{x}_3 &= bx_3\\ \dot{x}_4 &= c(0,0,x_3,x_4)\end{aligned}$$

is clearly unstable at the origin. Thus, NLSS is *not* solvable. However, assumptions of theorem 4.17 are satisfied. Moreover, the dynamics (4.7), associated to (5.1),

$$\dot{x}_4 = c(0,0,0,x_4)$$

is locally asymptotically stable at $x_4 = 0$ and each system (4.8), associated to (5.1),

$$\begin{aligned}\dot{x}_i &= u_i\\ \dot{x}_3 &= a_i(x_i) + bx_3\end{aligned} \tag{5.2}$$

is locally *exponentially* stabilizable at the origin via static feedback. Following the constructive procedure of section 4.2, let $n^w = 4, n_0 = 1, n_i = 1$, $i = 1,2$, and define

$$X_{11}^e(x^e) = \begin{pmatrix}1\\0\\0\\1\\0\\0\\0\end{pmatrix}, \quad X_{12}^e(x^e) = \begin{pmatrix}0\\0\\(\partial a_1(x_1)/\partial x_1)(x_1)\\0\\(\partial a_1(x_1)/\partial x_1)(x_1)\\0\\0\end{pmatrix},$$

$$X_{21}^e(x^e) = \begin{pmatrix}0\\1\\0\\0\\0\\1\\0\end{pmatrix}, \quad X_{22}^e(x^e) = \begin{pmatrix}0\\0\\(\partial a_2(x_2)/\partial x_2)(x_2)\\0\\0\\0\\(\partial a_2(x_2)/\partial x_2)(x_2)\end{pmatrix},$$

and $\mathcal{R}_i^e = \text{span}\{X_{ij}^e : j = 1,2\}$. It is worth noting that $X_{ik}^e(x^e)$ depends only on x_i. From remark 4.14, a feedback law which satisfies (4.31) and (4.32) is given by

$$\alpha^e(x^e) = \begin{pmatrix}0 & 0 & 0 & (a_1(x_1) + b\mu_1) & 0 & (a_2(x_2) + b\mu_2)\end{pmatrix}^T,$$

$$\beta^e(x^e) = \begin{pmatrix}1&0&0&0&0&0\\0&1&0&0&0&0\\1&0&1&0&0&0\\0&0&0&1&0&0\\0&1&0&0&1&0\\0&0&0&0&0&1\end{pmatrix}$$

The closed-loop system (4.51) can be transformed into (4.53) by means of the *linear* change of coordinates

$$\begin{aligned} z_1^e &= x_1 \\ z_2^e &= \mu_1 \\ z_3^e &= x_2 \\ z_4^e &= \mu_2 \\ z_5^e &= \lambda_1 - x_1 \\ z_6^e &= \lambda_2 - x_2 \\ z_7^e &= x_3 - \mu_1 - \mu_2. \end{aligned}$$

The system (4.8) is locally diffeomorphic to the system (4.54)

$$\begin{aligned} \dot{x}_i &= u_i \\ \dot{\mu}_i &= a_i(x_i) + b\mu_i. \end{aligned}$$

Thus, a feedback law which solves NLSD is given by

$$\begin{aligned} u_i &= H_{i1}x_i + H_{i3}\mu_i, \qquad i = 1,2, \\ \dot{\lambda}_i &= H_{i1}x_i + H_{i3}\mu_i - \lambda_i + x_i, \qquad i = 1,2, \\ \dot{\mu}_1 &= a_1(x_1) - bx_3 - b\mu_2, \\ \dot{\mu}_2 &= a_2(x_2) - bx_3 - b\mu_1, \end{aligned}$$

where H_{i1} and H_{i3} are real numbers such that the matrix

$$\begin{pmatrix} H_{i1} & H_{i3} \\ \frac{\partial a_i}{\partial x_i}(0) & b \end{pmatrix}$$

is Hurwitz. If we remove the assumption $(\partial a_i/\partial x_i)(0) \neq 0$, some system (5.2) may, in general, not be locally asymptotically stabilizable at the origin, even by means of continuous feedback laws (see [22] and [23] for an extensive analysis of all the possible cases); for example, this is the case when $a_i(x_i) = x_i^k$, with k even and $k \geq 2$. In general, if each system (5.2) is locally asymptotically stabilizable at the origin via C^k-feedback laws, then there exists a C^k-feedback law which solves NLSD for (5.1).

The dimension n^w of the dynamic feedback law, proposed above to solve NLSD for (5.1), can be reduced to 1. As a matter of fact, since $p = 2$, assumption (R3) is trivially satisfied. Moreover, it can be easily checked that also assumption (R2) holds. According to the results contained in section 4.3, pick $n^w = \dim(\mathcal{R}_1^* \cap \mathcal{R}_2^*) = 1$ with

$$X_{11}^e(x^e) = \begin{pmatrix} 1 \\ 0 \\ 0 \\ 0 \end{pmatrix}, \; X_{12}^e(x^e) = \begin{pmatrix} 0 \\ 0 \\ (\partial a_1(x_1)/\partial x_1)(x_1) \\ 0 \end{pmatrix},$$

$$X_{21}^e(x^e) = \begin{pmatrix} 0 \\ 1 \\ 0 \\ 0 \end{pmatrix}, \quad X_{22}^e(x^e) = \begin{pmatrix} 0 \\ 0 \\ (\partial a_2(x_2)/\partial x_2)(x_2) \\ (\partial a_2(x_2)/\partial x_2)(x_2) \end{pmatrix}.$$

It is easy to construct examples for which neither NLSS nor NLSD are solvable. The following system

$$\begin{aligned} \dot{x}_1 &= u_1, \\ \dot{x}_2 &= u_2, \\ \dot{x}_3 &= x_1 + x_2 + x_3^2, \\ y_i &= x_i, \qquad i = 1, 2, \end{aligned} \tag{5.3}$$

does *not* satisfy the conditions of either theorem 3.5 or theorem 4.22. Indeed, as it can be checked by straightforward computations, in a neighbourhood of x_0 we have

$$\begin{aligned} \mathcal{R}_i^* &= \text{span}\{\frac{\partial}{\partial x_i}, \frac{\partial}{\partial x_3}\}, \qquad i = 1, 2, \\ \Delta_{MIX}^* &= \mathcal{R}^* = \text{span}\{\frac{\partial}{\partial x_3}\}, \end{aligned}$$

and the dynamics (3.5) and (4.7) (which, in this case, coincide) are unstable at the origin.

Now, let us consider the following system (see also [92])

$$\begin{aligned} \dot{x}_1 &= x_3 + u_1, \\ \dot{x}_2 &= x_2 + u_1, \\ \dot{x}_3 &= u_2, \\ \dot{x}_4 &= a_1(x_1, x_3 + u_1) + a_2(x_2, x_2 + u_1) + bx_4, \\ \dot{x}_5 &= c(x_1, x_3 + u_1, x_2, x_2 + u_1, x_5), \\ y_i &= x_i, \qquad i = 1, 2, \end{aligned} \tag{5.4}$$

with $b > 0$. Assume that the dynamics $\dot{x}_5 = c(0,0,0,0,x_5)$ is locally asymptotically stable at $x_5 = 0$ and that each system

$$\begin{aligned} \dot{z}_{1i} &= z_{2i}, \\ \dot{z}_{2i} &= u_i, \\ \dot{z}_{3i} &= a_i(z_{1i}, z_{2i}) + bz_{3i}, \end{aligned}$$

is locally asymptotically stabilizable at the origin via dynamic feedback.

Apply the canonical dynamic extension algorithm to (5.4) (its decoupling matrix has rank 1 for all x). We have after the first step, with $p_C^0(x_C^0) = 1$ and $q_C^0(x_C^0) = 0$,

$$\begin{aligned}
\dot{x}_1 &= x_3 + \xi^0, \\
\dot{x}_2 &= x_2 + \xi^0, \\
\dot{x}_3 &= u_2, \\
\dot{x}_4 &= a_1(x_1, x_3 + \xi^0) + a_2(x_2, x_2 + \xi^0) + bx_4, \\
\dot{x}_5 &= c(x_1, x_3 + \xi^0, x_2, x_2 + \xi^0, x_5), \\
\dot{\xi}^0 &= u_{C1}^1, \\
y_i &= x_i, \qquad i = 1, 2,
\end{aligned}$$

which has vector relative degree at the origin of the state space. thus, we have obtained $\Sigma_C^{k_C^*}$ with $k_C^* = 1$. After the change of coordinates $z = (z_1(x), \ldots, z_6(x)) = (x_1, x_3 + \xi^0, x_2, x_2 + \xi^0, x_4, x_5)$ and the invertible static feedback transformation

$$\begin{pmatrix} u_{C1}^1 \\ u_2 \end{pmatrix} = \begin{pmatrix} 0 & 1 \\ 1 & -1 \end{pmatrix} \left\{ - \begin{pmatrix} 0 \\ z_4 \end{pmatrix} + \begin{pmatrix} v_1 \\ v_2 \end{pmatrix} \right\},$$

we obtain the following noninteractive system of the form (5.1)

$$\begin{aligned}
\dot{z}_1 &= z_2, \\
\dot{z}_2 &= v_1, \\
\dot{z}_3 &= z_3, \\
\dot{z}_4 &= v_2, \\
\dot{z}_5 &= a_1(z_1, z_2) + a_2(z_3, z_4) + bz_5, \\
\dot{z}_6 &= c(z_1, z_2, z_3, z_4, z_6), \\
y_1 &= z_1, \\
y_2 &= z_3.
\end{aligned}$$

Under the initial assumptions on (5.4) and reasoning as in the case of (5.1), it follows that there exists indeed a feedback law which solves NLSD for (5.4).

CHAPTER 4

NONINTERACTING CONTROL WITH GLOBAL STABILITY FOR NONLINEAR SYSTEMS: BASIC PRINCIPLES

4.1 Problem formulation

In the previous chapter, for the class of systems (**3**.1.1) we have studied the problem of achieving noninteracting control and stability through feedback laws defined in a neighbourhood of the origin of the state space. It is natural to ask under which additional assumptions these feedback laws are defined *globally* or, in other words, at each point of the state space. Behind this question, there is the major problem of rendering a set of distributions simultaneously invariant through feedback laws, which are defined globally. The theory of partial differential equation, which has been called upon in chapter **3** to solve this problem in a neighbourhood of the origin of the state space, has a *local* nature and, as such, it is not a natural tool in solving global questions.

Recently Dayawansa et al. ([20] and [21]) have pointed out the key role of the theory of connections and parallel transport of vector fields along leaves in finding a globally defined feedback law which renders invariant a given distribution. Indeed, parallel transport of vector fields and global asymptotic stabilization of interconnected systems are the basic ingredients we need in order to extend the results of chapter **2** to a global setting.

In this chapter, for the class of systems (**3**.1.1), we consider two problems, which are the natural extensions of NLSS and NLSD, respectively, to a global setting.

Noninteracting control with global stability via static feedback laws (NGSS). *Find, if possible, a static feedback law (**3**.1.2), defined for all $\mathbb{R}^n$, such that the system (**3**.1.1)–(**3**.1.2) is noninteractive, globally asymptotically stable in x_0 and has uniform vector relative degree.*

Noninteracting control with global stability via dynamic feedback laws (NGSD). *Find, if possible , $n^w \geq 0$ and a dynamic feedback law (**3**.1.2), defined for all $\mathbb{R}^{n+n^w}$, such that the system (**3**.1.1)–(**3**.1.2) is noninteractive, globally asymptotically stable in (x_0, w_0) and has uniform vector relative degree.*

Throughout this chapter, we assume that the state is available for measurement.

4.2 Some facts about parallel transport of vector fields along leaves of a constant dimensional and involutive distribution.

In this section, we want to give an intuitive idea of what parallel transport of vector fields is (for an exhaustive study of parallel transport, the reader is referred to the standard texts [7] and [56]) and how it can be related to the problem of achieving noninteraction.

Let us consider a smooth curve $s : t \to s(t), t_1 \leq t \leq t_2$, $s(t) \in I\!R^n$, and a smooth vector field $X \in I\!R^n$. If $X(s(t)) = \sum_{i=1}^{n} c_i(s(t))(\frac{\partial}{\partial x_i})(s(t))$, where $X(s(t))$ and $(\frac{\partial}{\partial x_i})(s(t))$ are the vectors of $I\!R^n$ assigned to the point $s(t)$ by the vector fields X and $\frac{\partial}{\partial x_i}$, respectively, we define the *derivative of* X *along* s *at* $s(t)$ as

$$\frac{dX}{dt}(s(t)) = \lim_{\Delta t \to 0} \frac{1}{\Delta t}[X(s(t+\Delta t)) - X(s(t))] = \sum_{i=1}^{m} \dot{c}_i(s(t)) \frac{\partial}{\partial x_i}(s(t)). \tag{2.1}$$

In (2.1), by $X(s(t+\Delta t)) - X(s(t))$ we mean the difference between the vector assigned to the point $s(t+\Delta t)$ by the vector field X and *translated back* to the point $s(t)$ and the vector assigned to $s(t)$ by X. We translate $X(s(\Delta t + t))$ by means of the standard translation of vectors in $I\!R^n$. Unfortunately, if, instead of $I\!R^n$, we consider any smooth manifold M (not diffeomorphic to $I\!R^n$), the above definition of derivative of vector fields (on M) along a curve (on M) does not work any more, since there is no *canonical* way to translate the vector $X(s(\Delta t + t))$ back to the point $s(t)$. A *connection* $\frac{D}{dt}$ is a way of differentiating a vector field X (on M) along a curve s (on M) and $\frac{DX}{dt}$ is the smooth vector field (on M) resulting from such differentation. We will say that X is *parallel along* s (with respect to a given connection $\frac{D}{dt}$) if $(\frac{DX}{dt})(s(t)) = 0$ for all $t_1 \leq t \leq t_2$.

It is not hard to show that there exists a *unique* smooth vector field X, defined on a given smooth curve $s \in M$ and parallel along s, such that $X(s(t_1)) = X_0$, with X_0 *a priori* given. The same holds true for a set of vector fields, which are linearly independent at $s(t_1)$ (*frames*). This carrying parallel vector fields (or frames) along curves is known as *parallel transport* (along s) and allows to translate vector fields (or frames) along curves in a *canonical* way. Anyway, this kind of translation depends, in general, on the curve chosen. In $I\!R^n$, parallel transport is *independent* of such a curve and depends only on the initial and final points of the curve.

Now, let $\Delta \subset I\!R^n$ be an involutive and constant dimensional distribution. Moreover, assume that

$$[g_j, \Delta] \subset \Delta + \mathcal{G}, \qquad j = 1, \dots, m. \tag{2.2}$$

It is well known that (2.2) is necessary and sufficient to find an $m \times m$ smooth matrix $\beta(x) = (\, \beta_1(x) \quad \cdots \quad \beta_m(x)\,)$, defined and invertible for all x in a neighbourhood of x_0 and such that

$$[G\beta_j, \Delta] \subset \Delta, \qquad j = 1, \ldots, m. \tag{2.3}$$

This is not true, in general, if we require $\beta(x)$ to be defined and invertible for all $x \in \mathbb{R}^n$. In this case, as it will be seen, there is a nontrivial topological obstruction: for this reason, following [21], we will say that a distribution $\Delta \subset \mathbb{R}^n$, which satisfies (2.2), is *weakly controlled invariant* for (**3**.1.1).

The inclusion (2.2) has a precise meaning in terms of parallel transport. To see this, it is sufficient to restrict our analysis to a neighbourhood U_p of an arbitrary point $p \in \mathcal{L}_x^\Delta$. The corresponding analysis in a global framework goes in the same way, but requires some fundamental concepts from the theory of connections on vector bundles, which go beyond the purposes of these notes (the interested reader is referred to [9agg] and [18]). By Frobenius' theorem, there exists a change of coordinates $\eta(x) = (\,\eta_1^T(x) \quad \eta_2^T(x)\,)^T$, $\eta(x_0) = 0$, defined in a neigbourhood $U_p' \subset U_p$ of p and such that (**2**.1.1) has the form

$$\dot\eta_1 = f_1(\eta_1, \eta_2) + \sum_{j=1}^{m} g_{1j}(\eta_1, \eta_2) u_j,$$

$$\dot\eta_2 = f_2(\eta_1, \eta_2) + \sum_{j=1}^{m} g_{2j}(\eta_1, \eta_2) u_j,$$

with $\Delta = \mathrm{span}\{\partial/\partial\eta_1\}$. Moreover, we will assume that $\Delta \cap \mathcal{G} = 0$ (see [DCBT] for a more general setting).

Let us define the following *parallelism* along leaves of Δ. Let $s : t \to s(t)$, $0 \le t \le t_1$, be a smooth curve of $\mathcal{L}_x^\Delta \cap U_p'$, q be the dimension of $\Delta(x)$ and $\bar Y$ be a smooth vector field, defined on U_p', which assigns to each point (η_1, η_2) a vector of $\mathbb{R}^{n-q}$ (thought of as immersed canonically in $\mathbb{R}^n$). We say that $\bar Y$ is *parallel* along s if $\bar Y(s(0)) = \bar Y(s(t))$ for all $0 \le t \le t_1$. This parallelism is independent of the curve chosen in $\mathcal{L}_x^\Delta \cap U_p'$ and can be extended to any smooth curve of $\mathcal{L}_x^\Delta$ (*natural parallelism along leaves* of Δ) This parallelism can be defined through a connection $\frac{D}{dt}$ such that $D\bar Y/dt$ denotes the last $n-q$ components of $[\bar Y, X]$), with $X \in \Delta$ such that $X(s(t)) = \dot s(t)$. It can be shown that $\bar Y$ is parallel (with respect to the connection $\frac{D}{dt}$) if and only if the last $n-q$ components of $[\bar Y, X](\eta_1, \eta_2)$ are identically zero for all $(\eta_1, \eta_2) \in U_p'$. It is easy to see that the vector fields

$$\bar g_j(\eta_2) = \begin{pmatrix} 0_{q\times 1} \\ g_{2j}(\eta_1, \eta_2)\big|_{\eta_1 = \eta_{1p}} \end{pmatrix}, \qquad j = 1, \ldots, m,$$

where (η_{1p}, η_{2p}) denotes the point p in η–coordinates, are parallel along leaves of Δ. For simplicity of notation, in what follows, we we will think of a vector $d \in \mathbb{R}^{n-q}$ as immersed canonically in $\mathbb{R}^n$.

Now, project the subspace $(\mathcal{G} + \Delta)(\eta_1, \eta_2)$ along $\Delta(\eta_1, \eta_2)$ or, equivalently, consider the subspace of $\mathbb{R}^{n-q}$ spanned by the vectors $g_{21}(\eta_1, \eta_2), \ldots, g_{2m}(\eta_1, \eta_2)$. It is easy to see that, if (2.2) holds, then at each point $(\eta_1, \eta_2) \in \mathcal{L}_x^\Delta \cap U_p'$ the subspace spanned by $g_{21}(\eta_1, \eta_2), \ldots, g_{2m}(\eta_1, \eta_2)$ is *independent* of the point (η_1, η_2) (see [18] for a proof in terms of bundles and connections). We will say in this case that span$\{g_{21}, \ldots, g_{2m}\}$ is *invariant under parallel transport* (on $\mathcal{L}_x^\Delta \cap U_p'$). Indeed, let q_1 and q_2 be two points of $\mathcal{L}_x^\Delta \cap U_p'$ and s be a smooth curve on $\mathcal{L}_x^\Delta \cap U_p'$, which joins q_1 to q_2. Since $\Delta \cap \mathcal{G} = 0$, at each $(\eta_1, \eta_2) \in U_p'$ the vectors $g_{21}(\eta_1, \eta_2), \ldots, g_{2m}(\eta_1, \eta_2)$ are linearly independent. Now, choose vectors $\bar{g}_{m+1}(\eta_2), \ldots, \bar{g}_{n-q}(\eta_2)$ such that the matrix

$$(\bar{g}_1(\eta_2) \quad \cdots \quad \bar{g}_m(\eta_2) \quad \bar{g}_{m+1}(\eta_2) \quad \cdots \quad \bar{g}_{n-q}(\eta_2))$$

is invertible for all η_2 in a neighbourhood of η_{2p}. As a consequence, we have $g_{2j}(\eta_1, \eta_2) = \sum_{k=1}^{n-q} \gamma_{jk}(\eta_1, \eta_2)\bar{g}_k(\eta_2)$, $j = 1, \ldots, m$, with $\gamma_{jk}(\eta_{1p}, \eta_{2p}) = 0$ for $k > m$. Since $\bar{g}_1, \ldots, \bar{g}_{n-q}$ are parallel along leaves of Δ or, equivalently, $\frac{D\bar{g}_j}{dt}(s(t)) = 0$ for $j = 1, \ldots, n-q$, we have

$$\frac{Dg_{2j}}{dt}(s(t)) = \sum_{k=1}^{n-q} \dot{\gamma}_{jk}(s(t))\bar{g}_k(s(t)), \qquad j = 1, \ldots, m.$$

Since $[g_j, X] \subset \Delta + \mathcal{G}$ for $X \in \Delta$, $\Delta \cap \mathcal{G} = 0$ and, moreover, $\frac{D\bar{g}_{2j}}{dt}$ denotes the last $n-q$ components of $[g_j, X]$), we obtain

$$\sum_{k=1}^{n-q} \dot{\gamma}_{jk}(s(t))\bar{g}_k(s(t)) = \sum_{h=1}^{m} \sum_{k=1}^{n-q} \theta_{jh}(s(t))\gamma_{hk}(s(t))\bar{g}_k(s(t)),$$
$$j = 1, \ldots, m, \; k = 1, \ldots, n-q,$$

where the θ_{jh}'s are defined by $\frac{D\bar{g}_j}{dt}(s(t)) = \sum_{h=1}^{m} \theta_{jh} g_{2h}$. Thus, it must be

$$\dot{\gamma}_{jk}(s(t)) = \sum_{h=1}^{m} \theta_{jh}(s(t))\gamma_{hk}(s(t)), \qquad j = 1, \ldots, m, \; k = 1, \ldots, n-q.$$

Since $\gamma_{jk}(\eta_{1p}, \eta_{2p}) = 0$ for $k > m$, it follows that $\gamma_{jk}(\eta_1, \eta_2) = 0$ for $k > m$ and for all $(\eta_1, \eta_2) \in s(t)$, which proves the desired result.

An immediate consequence of the above facts is that there exists a smooth matrix $\beta(\eta) = (\beta_1(\eta) \quad \cdots \quad \beta_m(\eta))$, defined and invertible for all $\eta \in U_p'$ and such that

$$(\bar{g}_1(\eta_2) \quad \cdots \quad \bar{g}_m(\eta_2)) = (g_{21}(\eta) \quad \cdots \quad g_{2m}(\eta))\,\beta(\eta).$$

The vector fields $G\beta_j$, $j = 1, \ldots, m$, satisfy (2.3) and the matrix $\beta(\eta)$ is defined in a neighbourhood of p. If η is a global diffeomorphism of $\mathbb{R}^n$, $\beta(\eta)$ is defined for *all* $\eta \in \mathbb{R}^n$ (see [21] for an extensive study in terms of bundles and basic connections).

We sum up the above results in the following proposition, which will be of much help in the next section.

Proposition 2.1. Let $\eta(x) = (\eta_1^T(x) \quad \eta_2^T(x))^T$, $\eta(x_0) = 0$, *be a global diffeomorphism of* $\mathbb{R}^n$ *such that* $\Delta = \mathrm{span}\{\partial/\partial\eta_1\}$. *Moreover, suppose that* Δ *has constant dimension for all* $x \in \mathbb{R}^n$ *and* $\Delta \cap \mathcal{G} = 0$. *Then, there exists a matrix* $\beta(x) = (\beta_1(x) \quad \cdots \quad \beta_m(x))$, *defined and invertible for all* $x \in \mathbb{R}^n$, *which satisfies (2.3).* □

4.3 Noninteracting control with global stability via static feedback

In what follows, we will denote by *globally invertible* feedback law any feedback law such that $M_{k^*} = \mathbb{R}^{n+n^w} \times \mathbb{R}^{mk^*}$ (see section **3**.1). Let us consider the class of globally invertible static feedback laws and assume the following.

*(A5) The system (**3**.1.1) has uniform vector relative degree.*

Moreover, we will assume that (A2), (A3) and (A4) hold for all $x \in \mathbb{R}^n$.

Under assumption (A5), the feedback law (**3**.3.1) is defined for all $x \in \mathbb{R}^n$. However, there does not exist, in general, a change of coordinates, defined for all $x \in \mathbb{R}^n$ and such that the system (**3**.1.1)–(**3**.3.1), has in these coordinates the form (**3**.3.8). The following theorem gives necessary and sufficient condition for the existence of such a change of coordinates. We recall that a vector field X is said to be *complete* if its flow $\Phi_t^X(p)$ is defined for all $t \in \mathbb{R}$ and for all $p \in \mathbb{R}^n$.

Proposition 3.1. There exists a set of smooth vector fields $X_{ji} \in \mathbb{R}^n$, $j = 1, \ldots, m+1$, $i = 1, \ldots, t_j$,

a) linearly independent at each $x \in \mathbb{R}^n$,

b) complete

and such that

c) $[X_{ji}, X_{kh}](x) = 0$ *for all* $x \in \mathbb{R}^n$, $j, k = 1, \ldots, m+1$, $i = 1, \ldots, t_j$ *and* $h = 1, \ldots, t_k$,

d) $\sum_{j \neq i} \mathcal{R}_j^* = \mathrm{span}\{X_{jh} : j \neq i,\, h = 1, \ldots, t_j\}$, *if and only if there exists a change of coordinates $z(x) = (\, z_1^T(x) \quad \cdots \quad z_{m+1}^T(x)\,)^T$, $z(x_0) = 0$, defined for all $x \in \mathbb{R}^n$ and such that $\widetilde{\Sigma}$ is expressed in these coordinates by (**3**.3.8).* □

Proof. The necessity is trivial. Following [19] (theorem 1), it is easy to show that, under assumptions a)–c), the map

$$\Psi : \mathbb{R}^n \times \mathbb{R}^n \to \mathbb{R}^n,$$

$$(z,x) \to \Psi(z,x) = \Phi_{z_{11}}^{X_{11}} \circ \cdots \circ \Phi_{z_{1t_1}}^{X_{1t_1}} \circ \cdots \circ \Phi_{z_{m+1,1}}^{X_{m+1,1}} \circ \cdots \circ \Phi_{z_{m+1,t_{m+1}}}^{X_{m+1,t_{m+1}}}(x),$$

where $\Phi_{z_{ij}}^{X_{ij}}$ denotes the flow associated to the vector field X_{ij}, is a global diffeomorphism for each fixed $x \in \mathbb{R}^n$. Moreover, if $\Psi_x = \Psi\big|_{\mathbb{R}^n \times \{x\}}$, it follows that

$$(\Psi_0)_*(\frac{\partial}{\partial z_{ji}}) = X_{ji}, \qquad j = 1, \ldots, m+1\,,\, i = 1, \ldots, t_j.$$

Thus $\Psi_0^{-1} : \mathbb{R}^n \to \mathbb{R}^n$ is the desired change of coordinates. □

It is important to note that the vector fields X_{ji} are *not* necessarily of the form (**3**.4.23).

Reasoning as in section **3**.3, we can prove the following result, which gives a necessary and sufficient condition to solve GNSS for the class of systems considered.

*Theorem 3.2. Let us consider the class of globally invertible static feedback laws and assume that (A2) and (A3) hold for all $x \in \mathbb{R}^n$. Moreover, assume that (A5) holds and (**3**.1.1) admits the global decomposition (**3**.3.8). Then, NGSS is solvable if*

*a) each system (**3**.3.10) is globally asymptotically stabilizable at the origin through a feedback law of the form*

$$u_i = \alpha_i(x_i), \qquad i = 1, \ldots, m.$$

*If, in addition, the distributions $\langle \widetilde{f}, \widetilde{g}_1, \ldots, \widetilde{g}_m | \mathrm{span}\{\widetilde{g}_i\}\rangle$, $i = 1, \ldots, m$, have constant dimension for all $x \in \mathbb{R}^n$, then a) holds if NGSS is solvable. In particular, the dynamics (**3**.3.9) is globally asymptotically stable at the origin and each system (**3**.3.10) is globally asymptotically stabilizable at the origin via static feedback.* □

Remark 3.3. The system (**3**.3.8) is *unique* up to global changes of coordinates and globally invertible static feedback transformations. As such, it can be uniquely associated to a given system (**3**.1.1) which satisfies the assumptions of proposition 3.2. Its stabilizability properties are independent of the feedback law chosen to obtain it. □

Remark 3.4. Condition a) of theorem 3.2 amounts to globally asymptotically stabilize (**3**.3.8) at the origin, while preserving its decoupled structure. The global stabilization of interconnected systems of the form (**3**.3.8) has been widely investigated in [11]–[12], [79]–[81] and [84]–[85]. □

4.4 Noninteracting control with global stability via dynamic feedback

In this section we will see how the results of section 4.2 can be used to solve NGSD for the class of systems (**3**.1.1), which satisfy assumptions (A2), (A3) and (A4) for all $x \in I\!R^n$, assumption (A5) and admit the global decompositions (**3**.4.4)–(**3**.4.5) and (**3**.4.21). We will not spend more words on these assumptions, since essentially the same remarks of chapter **2** hold.

Under the above assumptions and following section **3**.4.2, it is easy to show that, if NGSD is solvable, then the (globally defined) systems (**3**.4.7) and (**3**.4.8) must be globally asymptotically stable at the origin and, respectively, globally asymptotically stabilizable at the origin. The systems (**3**.4.7) and (**3**.4.8) are *uniquely* (up to global changes of coordinates and globally invertible static feedback transformations) associated to a given system (**3**.1.1), which satisfies assumptions (A2), (A3), (A4) for all $x \in I\!R^n$, assumption (A5) and admits the global decompositions (**3**.4.4)–(**3**.4.5) and (**3**.4.21).

Conversely, in the remaining part of the section we will prove that, if (**3**.4.7) is globally asymptotically stable at the origin, if each system (**3**.4.8) is globally asymptotically stabilizable at the origin and if, in addition, some convergent–input–convergent–state–like assumptions are satisfied, then NGSD is solvable.

For, assume that (**3**.4.7) is globally asymptotically stable at the origin and each system (**3**.4.8) is globally asymptotically stabilizable at the origin. If, in addition, we assume that the system

$$\dot{x}_{m+1,2} = \widetilde{f}_{m+1,2}(x_1,\ldots,x_m,x_{m+1,1},x_{m+1,2}) + \sum_{j=1}^{m} g_{m+1,2j}(x_1,\ldots,x_m,x_{m+1,1},x_{m+1,2})u_j \tag{4.1}$$

is convergent input convergent state, then, by reasoning as in section **3**.4.2 and from proposition **0**.2, it follows that, in order to solve NGSD for (**3**.3.7), we can assume without loss of generality that $\Delta_{MIX} = 0$. Moreover, in what follows we will assume that

*(A6) the vector fields X_{ik} of the form (**3**.4.23) are complete.*

This assumption is quite strong, in general, and ensures the existence of certain global decompositions (see the proof of the forthcoming lemma 4.1). Actually, as it will be clear later (lemma 4.6), it is sufficient to assume that there exist s_i vector fields $X_{i1}, \ldots, X_{is_i}$ of the form (**3**.4.23), complete and linearly independent at each $x \in I\!R^n$.

If we define

$$\varphi_i(x) = \begin{pmatrix} z_1^i(x) \\ z_3^i(x) \\ z_4^i(x) \end{pmatrix} , \; \bar{\varphi}_i(x) = z_2^i(x),$$

where $z_j^i(x)$ are defined as in section **3**.4.3, obviously we have $\mathcal{R}_i^* = \text{span}\{\partial/\partial\varphi_i\}$. If n_i and n_0 are the dimensions of the vectors x_i and $x_{m+1,1}$ respectively, let $\psi_i : I\!R^{n_i+n_0} \to I\!R^{s_i}$ and $\bar{\psi}_i : I\!R^{n_i+n_0} \to I\!R^{n_i+n_0-s_i}$ be the smooth mappings defined as

$$\begin{aligned} \psi_i(x_i, x_{m+1,1}) &= \varphi_i(x)\big|_{x_j=0 \text{ for } j \notin \{i,(m+1,1)\}} \quad , \\ \bar{\psi}_i(x_i, x_{m+1,1}) &= \bar{\varphi}_i(x)\big|_{x_j=0 \text{ for } j \notin \{i,(m+1,1)\}} \quad . \end{aligned} \tag{4.2}$$

Moreover, since by assumption $z^i(x)$ (see section **3**.4.3 for definition) is a global diffeomorphism of $I\!R^n$, it easily follows that x_i and $x_{m+1,1}$ can be uniquely expressed as smooth functions of ψ_i and $\bar{\psi}_i$ as

$$\begin{aligned} x_i &= \xi_i(\psi_i, \bar{\psi}_i), \\ x_{m+1,1} &= \bar{\xi}_i(\psi_i, \bar{\psi}_i). \end{aligned} \tag{4.3}$$

From (4.2) and (4.3), since $\widetilde{g}_i \in \bigcap_{j \neq i} \sum_{k \neq j} \mathcal{R}_k^*$ and $\widetilde{f}(x_0) = 0$, it is possible to define the following vector fields

$$\begin{aligned} f_i^*(\psi_i, \bar{\psi}_i) &= (\psi_i)_{*x}(\widetilde{f}\big|\mathcal{L}_{x_0}^{\bigcap_{j\neq i}\sum_{k\neq j}\mathcal{R}_k^*})(x)\Big|_{x_i=\xi_i(\psi_i,\bar{\psi}_i)\,,\,x_{m+1,1}=\bar{\xi}_i(\psi_i,\bar{\psi}_i)} \quad , \\ g_i^*(\psi_i, \bar{\psi}_i) &= (\psi_i)_{*x}(\widetilde{g}_i\big|\mathcal{L}_{x_0}^{\bigcap_{j\neq i}\sum_{k\neq j}\mathcal{R}_k^*})(x)\Big|_{x_i=\xi_i(\psi_i,\bar{\psi}_i)\,,\,x_{m+1,1}=\bar{\xi}_i(\psi_i,\bar{\psi}_i)} \quad , \end{aligned}$$

$$i = 1, \ldots, m.$$

Moreover, let

$$g_{m+i,k}^*(\psi_i, \bar{\psi}_i) = (\psi_i)_{*x}(T_{ik}\Big|\mathcal{L}_{x_0}^{\bigcap_{j\neq i}\sum_{k\neq j}\mathcal{R}_k^*})(x)\Big|_{x_i=\xi_i(\psi_i,\bar{\psi}_i)\,,\,x_{m+1,1}=\bar{\xi}_i(\psi_i,\bar{\psi}_i)} \quad ,$$

$$i = 1, \ldots, m \,,\, k = 1, \ldots, n_0,$$

where the T_{ik}'s are vector fields of $I\!R^n$, defined in the following lemma.

Lemma 4.1. Assume that (**3.1.1**) *satisfies assumptions (A2), (A3) and (A4) for all* $x \in \mathbb{R}^n$*, assumption (A5) and admits the global decompositions* (**3.4.4**)*–*(**3.4.5**) *and* (**3.4.21**)*. Then, there exist vector fields* T_{ik}*,* $i = 1, \dots, m$*,* $k = 1, \dots, n_0$*, defined and linearly independent at each* $x \in \mathbb{R}^n$ *and satisfying the following properties:*

a) $\mathrm{span}\{T_{i1}, \dots, T_{in_0}\} = \mathcal{R}^*$,

b) each component of $T_{ik}(x)$ *is a function only of* x_i *and* $x_{m+1,1}$,

c) $[X_{ih}, T_{ik}](x) = 0$ *for all* $x \in \mathcal{L}_{x_0}^{\bigcap_{j\neq i}\sum_{k\neq j}\mathcal{R}^*_k}$ *and for any vector field* X_{ih} *of the form* (**3.4.23**).

□

Remark 4.2. Lemma 4.1 does not follows from proposition 3.1, since the vector fields X_{ji} of proposition 3.1 do *not* satisfy, in general, either a) or b) of lemma 4.1. □

Proof (of lemma 4.1). The proof uses the results of section 4.2. Indeed, for each $i \in \{1, \dots, m\}$ let us consider the extended system (**2.4.25**) with $w = w_i = (\lambda_i^T \;\; \mu_i^T)^T$, $\lambda_i \in \mathbb{R}^{n_i}$ and $\mu_i \in \mathbb{R}^{n_0}$. Now, in a neighbourhood U_p of an arbitrary point $p \in \mathbb{R}^n$ consider a set of linearly independent vector fields $X_{i1}, \dots, X_{is_i}$ of the form (**2.4.23**) such that $\mathcal{R}_i^* = \mathrm{span}\{X_{i1}, \dots, X_{is_i}\}$ (note that we are not requiring that the vector fields $X_{i1}, \dots, X_{is_i}$ be linearly independent at *each* $x \in \mathbb{R}^n$). Correspondingly, define a distribution $\mathcal{R}_i^e \subset \mathbb{R}^{n+n^w}$ as follows. Let

$$X_{ik}^e(x^e) = \begin{pmatrix} X_{ik}(x) \\ X_{ik}^*(x_i, \mu_i) \end{pmatrix}, \qquad k = 1, \dots, s_i$$

where

$$X_{ik}^*(x_i, \mu_i) = \begin{pmatrix} Y_{ik}(x) \\ Z_{ik}(x) \end{pmatrix} \Big|_{x_j = 0 \text{ for } j \notin \{i,(m+1,1)\}\,;\, x_{m+1,1} = \mu_i},$$

and define

$$\mathcal{R}_i^e = \mathrm{span}\{X_{ik}^e : k = 1, \dots, s_i\}.$$

By reasoning as in remark **3**.4.9, it is easy to realize that $\mathcal{R}_i^e$ does *not* depend on the choice of the vector fields $X_{i1}, \dots, X_{is_i}$ and, thus, its definition is consistent for overlapping neighbourhoods of two different points $p, q \in \mathbb{R}^n$. As in section **3**.4.3, it can be proved that the distribution $\mathcal{R}_i^e$ is involutive and has constant dimension s_i. Moreover,

$$\begin{aligned} &[\mathcal{R}_i^e, \frac{\partial}{\partial \mu_i}] \subset \mathcal{R}_i^e + \mathrm{span}\{\frac{\partial}{\partial \mu_i}\}, \\ &\mathcal{R}_i^e \cap \mathcal{G}^w = 0. \end{aligned} \tag{4.4}$$

Suppose for a moment that there exists a global diffeomorphism $\eta_i : \mathbb{R}^{n+n^w} \to \mathbb{R}^{n+n^w}$, $\eta_i(x^e) = (\eta_{i1}^T(x^e) \;\; \eta_{i2}^T(x^e))^T$, $\eta_i(x_0^e) = 0$, such that $\mathcal{R}_i^e = \mathrm{span}\{\partial/\partial\eta_{i1}\}$. From theorem

2.1 and (4.4) it follows that there exists a $n^w \times n^w$ matrix $\beta^e(x^e)$, with smooth entries, defined and invertible at each $x^e \in \mathbb{R}^{n+n^w}$, such that

$$[X_{ik}^e, (\frac{\partial}{\partial \mu_i})\beta^e] \in \mathcal{R}_i^e, \tag{4.5}$$

for all $X_{ik}^e \in \mathcal{R}_i^e$. Moreover, by using lemma **3**.4.13, it is not hard to show that $\beta^e(x^e)$ can be chosen to be a function only of x_i and μ_i. Since $\mathcal{R}_i^e \cap \mathcal{G}^w = 0$ and from (4.5), it follows in particular that $[X_{ik}^e, (\frac{\partial}{\partial \mu_i})\beta^e](x^e) = 0$ for all $x^e \in \mathbb{R}^{n+n^w}$. By straightforward computations, if we define

$$\beta(x_i, x_{m+1,1}) = \beta^e(x^e)\Big|_{\mu_i = x_{m+1,1}} ,$$

it can be checked that $[X_{ik}, \frac{\partial}{\partial x_{m+1,1}}\beta](x) = 0$ for $x \in \mathcal{L}_{x_0}^{\bigcap_{j\neq i}\sum_{k\neq j}\mathcal{R}_k^*}$. Thus, we can define $T_{ik}(x)$ as the k–th column of the matrix

$$\begin{pmatrix} 0_{(n-n_O)\times n_O} \\ \beta(x_i, x_{m+1,1}) \end{pmatrix}.$$

The proof of the lemma is complete if we show that there exists a global diffeomorphism $\eta_i(x^e) = (\, \eta_{i1}^T(x^e) \quad \eta_{i2}^T(x^e)\,)^T$, with $\eta_i(x_0^e) = 0$, such that $\mathcal{R}_i^e = \text{span}\{\partial/\partial\eta_{i1}\}$.

To this end, define the following mapping (depending on i)

$$\delta : \mathbb{R}^{n+n^w} \to \mathbb{R}^{n+n^w},$$

$$x^e = \begin{pmatrix} x_1 \\ \vdots \\ x_m \\ x_{m+1,1} \\ \\ \lambda_i \\ \mu_i \end{pmatrix} \mapsto \begin{pmatrix} \bar{\psi}_i^*(x_i,\mu_i) \\ \psi_i^*(x_i,\mu_i) \\ x_1 \\ \vdots \\ x_{i-1} \\ x_{i+1} \\ \vdots \\ x_m \\ \lambda_i - x_i \\ x_{m+1,1} \end{pmatrix}, \tag{4.6}$$

where $\psi_i^* : \mathbb{R}^{n_i+n_O} \to \mathbb{R}^{s_i}$ and $\bar{\psi}_i^* : \mathbb{R}^{n_i+n_O} \to \mathbb{R}^{n_i+n_O} \to \mathbb{R}^{n_i+n_O-s_i}$ are smooth mappings defined as

$$\psi_i^*(x^e) = \psi_i(x)\Big|_{x_{m+1,1}=\mu_i},$$

$$\bar{\psi}_i^*(x^e) = \bar{\psi}_i(x)\Big|_{x_{m+1,1}=\mu_i}.$$

By direct inspection of (4.6), it follows that δ is a global diffeomorphism of $\mathbb{R}^{n+n^w}$. Moreover, let σ_i be the projection

$$\begin{aligned} \sigma_i : \mathbb{R}^{n+n^w} &\to \mathbb{R}^{s_i}, \\ x^e &\mapsto \sigma_i(x^e) = \psi_i^*(x_i, \mu_i). \end{aligned} \tag{4.7}$$

It can be easily seen that the restriction of σ_i to $\mathcal{L}_{x_0^e}^{\mathcal{R}_i^e}$ is a diffeomorphism onto. Indeed, reasoning as in the proof of proposition **3**.4.16, we conclude that $x_{m+1,1} = \mu_i$ and $x_i = \lambda_i$ on $\mathcal{L}_{x_0^e}^{\mathcal{R}_i^e}$. This, together with the definition of ψ_i^* and $\bar{\psi}_i^*$, implies our thesis.

In the same way, it can be proved that also the restriction of σ_i to $\mathcal{L}_{x^e}^{\mathcal{R}_i^e}$, where x^e is *any* point of $\mathbb{R}^{n+n^w}$, which satisfies the following constraints

$$\begin{aligned} x_j &= 0, \qquad j \notin \{i, (m+1,1)\}, \\ x_i &= \lambda_i, \\ x_{m+1,1} &= \mu_i, \end{aligned}$$

is a diffeomorphism onto. Now, consider an arbitrary point $\widetilde{x}^e \in \mathbb{R}^{n+n^w}$

$$\widetilde{x}^e = (\widetilde{x}_1, \ldots, \widetilde{x}_{i-1}, \widetilde{x}_i, \widetilde{x}_{i+1}, \ldots, \widetilde{x}_{m+1,1}, \widetilde{\lambda}_i, \widetilde{\mu}_i)$$

and, correspondingly, define

$$\bar{x}^e = (0, \ldots, 0, \widetilde{x}_i, 0, \ldots, \widetilde{\mu}_i, \widetilde{\lambda}_i, \widetilde{\mu}_i) \ .$$

As already pointed out above, the map $\bar{\sigma}_i : \sigma_i|\mathcal{L}_{\bar{x}^e}^{\mathcal{R}_i^e} \to \mathbb{R}^{s_i}$ is a diffeomorphism onto. First, we will use this fact to show that $\widetilde{\sigma}_i : \sigma_i|\mathcal{L}_{\widetilde{x}^e}^{\mathcal{R}_i^e} \to \mathbb{R}^{s_i}$ is onto as follows. Let $\widehat{x}^e$ be the inverse image under $\bar{\sigma}_i$ of an arbitrary point of $\mathbb{R}^{s_i}$. By Chow's theorem (see [14] or [55], lemma **3**.1.10), we can join $\bar{x}^e$ to $\widehat{x}^e$ by means of a concatenation of integral curves of the vector fields X_{ik}^e, $k = 1, \ldots, s_i$. In particular, the differential equations

$$\begin{aligned} \dot{x}_i &= Y_{ik}, \\ \dot{\mu}_i &= Z_{ik}(0, \ldots, x_i, 0, \ldots, \mu_i), \end{aligned} \tag{4.14}$$

describe uniquely the components x_i and μ_i of such integral curves, once an initial condition is given. If we replace the initial condition $\bar{x}^e$ with $\widetilde{x}^e$, the differential equations (4.14) have the same initial conditions. Since the right–hand part of these equations depends only on x_i and μ_i and the vector fields X_{ik} are complete, the components x_i and μ_i will be left unchanged if we replace the initial condition $\bar{x}^e$ by $\widetilde{x}^e$. Since the image of σ_i depends only on x_i and μ_i, clearly there exists a point $\bar{\bar{x}}^e \in \mathcal{L}_{\widetilde{x}^e}^{\mathcal{R}_i^e}$ such that $\widetilde{\sigma}_i(\bar{\bar{x}}^e) = \bar{\sigma}_i(\widehat{x}^e)$. This exactly amounts to $\widetilde{\sigma}_i$ being surjective.

Next, we show that $\widetilde{\sigma}_i$ is a diffeomorphism by first showing that the map $\varrho_i = (\bar{\sigma}_i)^{-1} \circ \widetilde{\sigma}_i : \mathcal{L}_{\widetilde{x}^e}^{\mathcal{R}_i^e} \to \mathcal{L}_{\bar{x}^e}^{\mathcal{R}_i^e}$ is a covering map ([7], pp. 100–104). For, fix $q_0 \in \mathcal{L}_{\bar{x}^e}^{\mathcal{R}_i^e}$ and consider $q_j \in \varrho_i^{-1}(q_0) \subset \mathcal{L}_{\widetilde{x}^e}^{\mathcal{R}_i^e}$ for $j \in I$, where I is some index set. Let us introduce the following maps

$$\begin{aligned}
\delta_{i0} &: (-\epsilon, \epsilon)^{s_i} \to \mathcal{L}_{\bar{x}^e}^{\mathcal{R}_i^e}, \\
&(t_1, \ldots, t_{s_i}) \mapsto \Phi_{t_{s_i}}^{X_{is_i}^e} \circ \cdots \circ \Phi_{t_1}^{X_{i1}^e}(q_0), \\
\delta_{ij} &: (-\epsilon, \epsilon)^{s_i} \to \mathcal{L}_{\widetilde{x}^e}^{\mathcal{R}_i^e}, \qquad j \in I, \\
&(t_1, \ldots, t_{s_i}) \mapsto \Phi_{t_{s_i}}^{X_{is_i}^e} \circ \cdots \circ \Phi_{t_1}^{X_{i1}^e}(q_j)
\end{aligned}$$

for some $\epsilon > 0$. It is well-known ([7], theorem 4.12) that $\epsilon > 0$ can be chosen sufficiently small in such a way that δ_{i0} is a diffeomorphism onto its image.

Now, let $V_j = \delta_{ij}((-\epsilon, \epsilon)^{s_i})$. The vector fields $X_{i1}^e, \ldots, X_{is_i}^e$ can be constructed in such a way to be linearly independent at each $x^e \in V_0 \cup (\bigcup_{\alpha} V_\alpha)$. As a matter of fact, as it will be shown in a moment, the components x_i and μ_i of q_0 and q_j, $j \in I$, respectively, are equal, which implies that the vectors $X_{ik}^e(x^e)$ are linearly independent at each $x^e \in \cup_\alpha V_\alpha$, if they are such at each $x^e \in V_0$. Thus, it suffices to construct $X_{i1}^e, \ldots, X_{is_i}^e$ linearly independent at each $x^e \in V_0$. Note also that ψ_{i0} and ψ_{ij}, $j = 1, \ldots, m$, are defined for the *same* ϵ, which is possible by completeness of the vector fields X_{ik}, $k = 1, \ldots, s_i$.

We will prove that V_j is open in $\mathcal{L}_{\widetilde{x}^e}^{\mathcal{R}_i^e}$ and ϱ_i is a diffeomorphism of V_j onto V_0. Indeed, denote by $x_i(x^e)$ and $\mu_i(x^e)$ the components x_i and μ_i, respectively, of x^e. We claim that $\mu_i(q_0) = \mu_i(q_j)$ and $x_i(q_0) = x_i(q_j)$ for $j \in I$. As a matter of fact, by definition of q_j and ϱ_i, it follows that $\psi_i^*(q_0) = \psi_i^*(q_j)$ for $j \in I$. Moreover, $L_{X_{ik}^e}\bar{\psi}_i^*(x^e) = 0$ for all x^e, since $L_{X_{ik}}\bar{\varphi}_i(x) = 0$ for all $x \in I\!R^n$. Thus, $\bar{\psi}_i^*(q_0) = \bar{\psi}_i^*(q_j)$ for $j \in I$, since $\bar{\psi}_i^*$ is constant along leaves of $\mathcal{R}_i^e$ and $\bar{\psi}_i^*$ has the same value at $\widetilde{x}^e$ and $\bar{x}^e$. From (4.2), it follows our claim. We conclude that

$$\sigma_i \circ \delta_{i0}(t_1, \ldots, t_{s_i}) = \sigma_i \circ \delta_{ij}(t_1, \ldots, t_{s_i}) \qquad i = 1, \ldots, m$$

for all $(t_1, \ldots, t_{s_i}) \in (-\epsilon, \epsilon)^{s_i}$. Since $\bar{\sigma}_i$ and δ_{i0} are diffeomorphisms, so is δ_{ij} and V_j is open.

Moreover, $\varrho_i^{-1}(V_0) = \cup_{j \in I} V_j$. Indeed, suppose that there exists $p \in \varrho_i^{-1}(V_0)$ but $p \notin V_\alpha$ for any $\alpha \in I$. Thus, $\varrho_i(p) \in V_0$, which implies $\varrho_i(p) = \delta_{i0}(t_1, \ldots, t_{s_i})$ for some $(t_1, \ldots, t_{s_i}) \in (-\epsilon, \epsilon)^{s_i}$. Consider the point $q_\alpha = \Phi_{-t_1}^{X_{i1}^e} \circ \cdots \circ \Phi_{-t_{s_i}}^{X_{is_i}^e}(p)$. Clearly, $\psi_i^*(q_0) = \psi_i^*(q_\alpha)$ and, consequently, $\varrho_i(q_\alpha) = q_0$, which implies $p \in V_\alpha$. This clearly gives a contradiction.

Since the open sets V_j, $j \in I$, can be taken disjoint, by making V_0 smaller if necessary, it follows that ϱ_i is a covering map. Since $\mathcal{L}_{\bar{x}^e}^{\mathcal{R}_i^e}$ is simply connected (being diffeomorphic to $\mathbb{R}^{s_i}$), we conclude that ϱ_i and, thus, $\tilde{\sigma}_i$ is a diffeomorphism (see [7], pp. 286–292).

Since the restriction of σ_i to $\mathcal{L}_{x^e}^{\mathcal{R}_i^e}$ is a diffeomorphism onto, $\eta_{i1} = \sigma_i(x^e)$ is a set of global coordinates for $\mathcal{L}_{x^e}^{\mathcal{R}_i^e}$. As a consequence, the intersection between each leaf of $\mathcal{R}_i^e$ and the smooth manifold $\psi_i^*(x^e) = 0$ is unique. We choose this intersection as the coordinate η_{i2}. Correspondingly, we obtain a smooth (possibly vector-valued) function

$$\psi_0^*(\psi_i^*, \bar{\psi}_i^*, x_1, \ldots, x_{i-1}, x_{i+1}, \ldots, x_m, -x_i + \lambda_i, x_{m+1,1}),$$

such that its level sets describe the leaves of $\mathcal{R}_i^e$. Now, consider the map (depending on i)

$$\eta : \mathbb{R}^{n+n^w} \to \mathbb{R}^{n+n^w},$$

$$\begin{pmatrix} \psi_i^* \\ \bar{\psi}_i^* \\ x_1 \\ \vdots \\ x_{i-1} \\ x_{i+1} \\ \vdots \\ x_m \\ \lambda_i - x_i \\ x_{m+1,1} \end{pmatrix} \mapsto \begin{pmatrix} \psi_i^* \\ \psi_0^*(\psi_i^*, \bar{\psi}_i^*, x_1, \ldots, x_{i-1}, x_{i+1}, \ldots, x_m, -x_i + \lambda_i, x_{m+1,1}) \end{pmatrix}.$$

and, for simplicity, denote $(\bar{\psi}_i^*, x_1, \ldots, x_{i-1}, x_{i+1}, \ldots, x_m, \lambda_i - x_i, x_{m+1,1})$ by θ. If we show that

$$\eta_{*(\psi_i^*, \theta)} : T_{(\psi_i^*, \theta)} \mathbb{R}^{n+n^w} \to T_{(\psi_i^*, \psi_0^*)} \mathbb{R}^{n+n^w}$$

is an isomorphism for all (ψ_i^*, θ), it immediately follows that η is a diffeomorphism onto and the lemma is proved.

For, assume for simplicity that ψ_i^* and θ are real-valued functions (the general case goes in the same way). Given $q_2 = (\psi_i^*, \theta)$, let q_1 be the intersection between $\mathcal{L}_{q_2}^{\mathcal{R}_i^e}$ and the smooth manifold $\psi_i^* = 0$. Since σ_i is a diffeomorphism when restricted to $\mathcal{L}_{q_2}^{\mathcal{R}_i^e}$, it follows that $\eta_{*q_2} : \mathcal{R}_i^e(q_2) \to T_{\eta(q_2)} \mathbb{R}^{s_i}$ is an isomorphism. Let

$$e_{\psi_i^*} = \eta_{*q_2}^{-1}((\frac{\partial}{\partial \psi_i^*})(\eta(q_2))),$$

and e_θ be any vector linearly independent from $e_{\psi_i^*}$. Choose any smooth curve σ, $t \to \sigma(t)$, $t \geq 0$ and $\sigma(t) \in \mathbb{R}^{n+n^w}$, such that $\sigma(0) = q_2$, $\dot{\sigma}(0)$ is nonzero and parallel to the vector e_θ. By definition of tangent vector and η_*, we have

$$\eta_{*q_2} e_\theta = (\frac{d}{dt}(\eta \circ \sigma))\Big|_{t=0}.$$

Since by construction the last component of $(\frac{d}{dt}(\eta\circ\sigma))\big|_{t=0}$ is nonzero, $\eta_{*q_2}(\, e_{\psi_i^*} \quad e_\theta\,)$ is nonsingular. Since also $(\, e_{\psi_i^*} \quad e_\theta\,)$ is nonsingular, η_{*q_2} is an isomorphism. □

Now, assume that each system (**3**.4.8) is globally asymptotically stabilizable at the origin. Since $x_j = 0$, $j \notin \{i,(m+1,1)\}$, for $x \in \mathcal{L}_{x_0}^{\mathcal{R}_i^*}$ and $\bar{\varphi}_i(x) = 0$ for $x \in \mathcal{L}_{x_0}^{\mathcal{R}_i^*}$, the system

$$\dot{\psi}_i = f_i^*(\psi_i,0) + g_i^*(\psi_i,0)u_i \tag{4.8}$$

is diffeomorphic to (**3**.4.8). Let

$$\begin{aligned} u_i &= \eta_i(\psi_i,\xi_i), \\ \dot{\xi}_i &= \zeta_i(\psi_i,\xi_i), \end{aligned} \tag{4.9}$$

be a feedback law which globally asymptotically stabilizes (4.8) at the origin, with $\eta_i(0,0) = 0$ and $\zeta_i(0,0) = 0$. In what follows, we assume that also the system

$$\begin{aligned} \dot{\psi}_i &= f_i^*(\psi_i,\bar{\psi}_i) + g_i^*(\psi_i,\bar{\psi}_i)\eta_i(\psi_i,\xi_i) + \sum_{j=m+1}^{2m} g_j^*(\psi_i,\bar{\psi}_i)u_j, \\ \dot{\xi}_i &= \zeta_i(\psi_i,\xi_i), \end{aligned} \tag{4.10}$$

is convergent input convergent state. Note that this property may depend on the feedback law (4.9) and on the vector fields $g_j^*(\psi_i,\bar{\psi}_i)$, $j = 1,\ldots,2m$ (or, wihich is the same, on the vector fields T_{ik}'s).

To prove that the above assumptions are sufficient to solve NGSD, we proceed according to the same strategy used to solve NLSD.

The distributions $\mathcal{R}_1^e,\ldots,\mathcal{R}_m^e$ are defined as follows. Given a point $\hat{x} \in I\!R^n$, choose vector fields $X_{i1},\ldots,X_{is_i}$ of the form (**3**.4.23), which are linearly independent at each x in a neighbourhood of $\hat{x}$ and span $\mathcal{R}_i^*$. Correspondingly, define $\mathcal{R}_i^e$ as in (**2**.4.27). The definition of $\mathcal{R}_i^e$ is consistent for overlapping neighbourhoods of two different points $\hat{x}_1,\hat{x}_2 \in I\!R^n$ (see the proof of lemma 4.1).

The distributions $\mathcal{R}_1^e,\ldots,\mathcal{R}_m^e$ are involutive and controlled invariant for (**3**.4.25). The following proposition extends proposition **3**.4.13 to a global setting.

Proposition 4.3. There exists a globally invertible feedback law $u^e = \alpha^e(x^e) + \sum_{j=1}^{2m}\beta_j^e(x^e)v_j^e$, *with* $\alpha^e(x_0^e) = 0$, *such that*

$$[\tilde{f}^e,\mathcal{R}_i^e] \subset \mathcal{R}_i^e, \qquad i = 1,\ldots,m, \tag{4.11}$$

$$[\tilde{g}_j^e,\mathcal{R}_i^e] \subset \mathcal{R}_i^e, \qquad j = 1,\ldots,2m,\ i = 1,\ldots,m, \tag{4.12}$$

and $\tilde{g}_i^e = X_{i1}^e$ *for* $i = 1,\ldots,m$. □

Proof. An explicit expression of $\alpha^e(x^e)$ is given by

$$\alpha^e(x^e) = \begin{pmatrix} 0_{m\times 1} \\ \tilde{f}_1(x_1) \\ \tilde{f}_{m+1,1}(x)\Big|_{x_j=0 \text{ for } j\notin\{1,(m+1,1)\},\, x_{m+1,1}=\mu_1} \\ \vdots \\ \tilde{f}_m(x_m) \\ \tilde{f}_{m+1,1}(x)\Big|_{x_j=0 \text{ for } j\notin\{m,(m+1,1)\},\, x_{m+1,1}=\mu_m} \end{pmatrix}$$

(see remark **3**.4.14). Moreover, it can be easily seen that, if

$$\beta_i^*(x_i,\mu_i) = \beta_i(x_i, x_{m+1,1})\Big|_{x_{m+1,1}=\mu_i} \quad ,$$

where $\beta_i(x_i, x_{m+1,1})$ is defined as in the proof of lemma 4.1, the following vector fields

$$\tilde{g}^e_{m+j} = \begin{pmatrix} 0_{n\times n_j} & 0_{n\times n_0} \\ 0_{(n_1+n_0)\times n_j} & 0_{(n_1+n_0)\times n_0} \\ \vdots & \vdots \\ 0_{(n_{j-1}+n_0)\times n_j} & 0_{(n_{j-1}+n_0)\times n_0} \\ I_{n_j\times n_j} & 0_{n_j\times n_0} \\ 0_{n_0\times n_j} & \beta_j^*(x_j,\mu_j) \\ 0_{(n_{j+1}+n_0)\times n_j} & 0_{(n_{j+1}+n_0)\times n_0} \\ \vdots & \vdots \\ 0_{(n_m+n_0)\times n_j} & 0_{(n_m+n_0)\times n_0} \end{pmatrix}, \qquad j=1,\ldots,m, \tag{4.13}$$

satisfy (4.12). From (4.13) we obtain the expression of $\beta^e_{m+1}(x^e),\ldots,\beta^e_{2m}(x^e)$. On the other hand, $\beta_I(x^e)$, $i=1,\ldots,m$, can be chosen as in proposition **3**.4.13. Moreover, the matrix $(\,\beta_1^e(x^e) \;\cdots\; \beta^e_{2m}(x^e)\,)$ is invertible for all $x^e\in\mathbb{R}^{n+n^w}$ by construction. □

By plugging the feedback law $u^e = \alpha^e(x^e) + \sum_{j=1}^{2m}\beta_j^e(x^e)v_j^e$ into (**3**.4.25), we obtain (**3**.4.51).

*Proposition 4.4. There exists a feedback law of the form (**2**.4.52), defined for all (x^e, η_i^e) and such that the system (**3**.4.51)-(**3**.4.52) is globally asymptotically stable at the origin, noninteractive and has uniform vector relative degree.* □

Proof. First, we will show that there exists a change of coordinates

$$z^e = (\,(z_1^e)^T(x^e) \;\cdots\; (z^e_{m+1})^T(x)\,)^T,$$

with $z^e(x_0^e) = 0$, defined for all $x^e \in \mathbb{R}^{n+n^w}$ and such that (2.4.51) is expressed in z^e–coordinates by (3.4.53). For, define the following mapping

$$\delta : \mathbb{R}^{n+n^w} \to \mathbb{R}^{n+n^w},$$

$$x^e = \begin{pmatrix} x_1 \\ \vdots \\ x_m \\ x_{m+1,1} \\ \lambda_1 \\ \mu_1 \\ \vdots \\ \lambda_m \\ \mu_m \end{pmatrix} \mapsto \begin{pmatrix} \psi_1^*(x_1, \mu_1) \\ \vdots \\ \psi_m^*(x_m, \mu_m) \\ \bar{\psi}_1^*(x_1, \mu_1) \\ \vdots \\ \bar{\psi}_m^*(x_m, \mu_m) \\ \lambda_1 - x_1 \\ \vdots \\ \lambda_m - x_m \\ x_{m+1,1} \end{pmatrix},$$

where ψ_i^* and $\bar{\psi}_i^*$ are defined as in lemma 4.1. Reasoning as in the proof of lemma 4.1, we can show that there exists a global diffeomorphism $z^e(x^e) = \left((z_1^e)^T(x^e) \quad \cdots \quad z_{m+1}^e)^T(x^e) \right)^T$, $z^e(x_0^e) = 0$, such that $\mathcal{R}_i^e = \operatorname{span}\{\partial/\partial z_i^e\}$, $i = 1, \ldots, m$. From here, with proposition 0.2 in mind, we can proceed as in the proof of proposition 3.4.16, once we prove that for each system

$$\dot{z}_i^e = \widetilde{f}_i^e(z_i^e, z_{m+1}^e) + \widetilde{g}_{ii}^e(z_i^e, z_{m+1}^e)v_i^e + \sum_{j=m+1}^{2m} \widetilde{g}_{ij}^e(z_i^e, z_{m+1}^e)v_j^e \tag{4.15}$$

there exists a feedback law

$$\begin{aligned} v_i^e &= \eta_i(z_i^e, \xi_i), \\ \dot{\xi}_i &= \zeta_i(z_i^e, \xi_i) \end{aligned} \tag{4.16}$$

such that

$$\begin{aligned} \dot{z}_i^e &= \widetilde{f}_i^e(z_i^e, 0) + \widetilde{g}_{ii}^e(z_i^e, 0)\eta_i(z_i^e, \xi_i), \\ \dot{\xi}_i &= \zeta_i(z_i^e, \xi_i), \end{aligned} \tag{4.17}$$

is globally asymptotically asymptotically stable at the origin and

$$\begin{aligned} \dot{z}_i^e &= \widetilde{f}_i^e(z_i^e, z_{m+1}^e) + \widetilde{g}_{ii}^e(z_i^e, z_{m+1}^e)\eta_i(z_i^e, \xi_i) + \sum_{j=m+1}^{2m} \widetilde{g}_{ij}^e(z_i^e, z_{m+1}^e)v_j^e, \\ \dot{\xi}_i &= \zeta_i(z_i^e, \xi_i) \end{aligned} \tag{4.18}$$

is convergent input convergent state.

Indeed, note that the coordinates $\bar{\psi}_1^*, \ldots, \bar{\psi}_m^*, \lambda_1 - x_1, \ldots, \lambda_m - x_m$ can be chosen as part of the coordinates ψ_0^* (see proof of lemma 4.1), since $L_{X_{ik}^e}\bar{\psi}_j^*(x^e) = 0$ and

$L_{X^e_{ik}}\lambda_j(x^e) = L_{X^e_{ik}}x_j(x^e)$ for all x^e, $i,j = 1,\dots,m$ and for all $X^e_{ik} \in \mathcal{R}^e_i$. Moreover, let $\varphi^e = z^e(x^e)$. From the structure of $\alpha^e(x^e)$ and $\beta^e(x^e)$ (see proposition 4.3) and (4.3), it follows that (4.15) is given in explicit terms by

$$\begin{aligned}\dot{\psi}^*_i &= (\psi^*_i)_{*x^e}(\widetilde{f}^e(x^e))\Big|_{x^e=(z^e)^{-1}(\varphi^e)} + ((\psi^*_i)_{*x^e}(\widetilde{g}^e_i(x^e))\Big|_{x^e=(z^e)^{-1}(\varphi^e)})v^e_i + \\ &\quad + \sum_{j=m+1}^{2m} ((\psi^*_i)_{*x^e}(\widetilde{g}^e_{m+j}(x^e))\Big|_{x^e=(z^e)^{-1}(\varphi^e)})v^e_j = \\ &= f^*_i(\psi^*_i,\bar{\psi}^*_i) + g^*_{ii}(\psi^*_i,\bar{\psi}^*_i)v^e_i + \sum_{j=m+1}^{2m} g^*_{m+j}(\psi^*_i,\bar{\psi}^*_i)v^e_j.\end{aligned}$$

From the initial assumptions on (4.10), it follows that there exists indeed a feedback law (4.16) such that (4.17) is globally asymptotically stable at the origin and (4.18) is convergent input convergent state. □

We sum up all the above results in the following theorem.

*Theorem 4.5. Assume that (A2), (A3) and (A4) hold for all $x \in \mathbb{R}^n$. Moreover, assume that (A5) and (A6) hold and (**3**.1.1) admits the global decompositions (**3**.4.4)-(**3**.4.5) and (**3**.4.21). Then, NGSD is solvable if*

*a) the dynamics (**3**.4.7) is globally asymptotically stable at the origin,*

b) the system (4.1) is convergent input convergent state,

c) there exists a dynamic feedback law (4.9) which globally asymptotically stabilizes (4.8) at the origin and such that the corresponding system (4.10) is convergent input convergent state.

*On the other hand, let (**3**.1.2) be a feedback law which solves NGSD. If, in addition, the distributions $\mathcal{R}^e_i = \langle \widetilde{f}^e, \widetilde{g}^e_1, \dots, \widetilde{g}^e_m | \text{span}\{\widetilde{g}^e_i\}\rangle$, Δ^e_{MIX}, $i = 1,\dots,m$, are regularly computable at each $x^e \in \mathbb{R}^{n+n^w}$ and the distributions $\Delta^e_{MIX} + \mathcal{R}^e_i$, $i = 1,\dots,m$, have constant dimension for all $x \in \mathbb{R}^{n+n^w}$, then a) holds and each system (4.8) is globally asymptotically stabilizable at the origin via dynamic feedback.* □

Remark 4.6. It is worth noting that the global decomposition (**3**.4.21) is required only for proving the necessary part of theorem 4.5. Moreover, if assumption (A6) is *not satisfied*, yet NGSD might be solvable. Indeed, although some vector fields of the form (**3**.4.23) may happen not to be complete, the proof of proposition 4.4 can be carried through as long as there exists a set of vector fields $X_{i1}, \ldots, X_{is_i}$ of the form (**3**.4.23), which are complete and linearly independent for all $x \in \mathbb{R}^n$. □

Let us consider the following example. Consider the following noninteractive system

$$\begin{aligned} \dot{x}_i &= u_i, \qquad i = 1, 2, \\ \dot{x}_3 &= -x_1^3 - x_2^3 + x_3, \\ y_i &= x_i, \qquad i = 1, 2. \end{aligned} \tag{4.19}$$

We want to prove that NGSD is solvable for (4.19). Indeed,

$$\begin{aligned} \Delta_{MIX} &= 0, \\ \mathcal{R}_i^* &= \text{span}\{\frac{\partial}{\partial x_i}, \frac{\partial}{\partial x_3}\} = \text{span}\{\tilde{g}_i, ad_{\tilde{f}}^3 \tilde{g}_i\}, i = 1, 2, \end{aligned}$$

and each system (**3**.4.8)

$$\begin{aligned} \dot{x}_i &= u_i, \\ \dot{x}_3 &= -x_i^3 + x_3, \end{aligned} \tag{4.20}$$

is *globally* asymptotically stabilizable at the origin via *continuous* static state–feedback ([16]). Note that the linear approximation of (4.20) has an uncontrollable eigenvalue with positive real part. Moreover, the vector fields X_{ik} of the form (**2**.4.23) do not depend on x_3. Note also that there exists at least one of these vector fields, which is *not* complete (for example, $ad_{\tilde{f}} \tilde{g}_i$).

Let us define the distributions $\mathcal{R}_1^e$ and $\mathcal{R}_2^e$ as in (**3**.4.27) and (**2**.4.28). We obtain

$$\mathcal{R}_1^e = \text{span}\{\begin{pmatrix} 1 \\ 0 \\ 0 \\ 1 \\ 0 \\ 0 \\ 0 \end{pmatrix}, \begin{pmatrix} 0 \\ 0 \\ 1 \\ 0 \\ 1 \\ 0 \\ 0 \end{pmatrix}\}, \ \mathcal{R}_2^e = \text{span}\{\begin{pmatrix} 0 \\ 1 \\ 0 \\ 0 \\ 0 \\ 1 \\ 0 \end{pmatrix}, \begin{pmatrix} 0 \\ 0 \\ 1 \\ 0 \\ 0 \\ 0 \\ 1 \end{pmatrix}\}.$$

Now, choose $\alpha^e(x^e)$, $\beta^e(x^e)$ and $z^e(x^e)$ as in the case of example (**3**.5.1). The change of coordinates $z^e(x^e)$ is defined for all $x^e \in \mathbb{R}^{n+n^w}$. If $u^w = (u_{11}^w \ \ u_{12}^w \ \ u_{21}^w \ \ u_{22}^w)^T$, the system (**3**.4.53) is given in z^e–coordinates by

$$\begin{aligned}
\dot{z}_1^e &= u_1, \\
\dot{z}_2^e &= u_{12}^w - (z_1^e)^3 + z_2^e, \\
\dot{z}_3^e &= u_2, \\
\dot{z}_4^e &= u_{22}^w - (z_3^e)^3 + z_4^e, \\
\dot{z}_5^e &= u_{11}^w, \\
\dot{z}_6^e &= u_{21}^w, \\
\dot{z}_7^e &= z_7^e - u_{12}^w - u_{22}^w.
\end{aligned} \tag{4.21}$$

Let $F_1(z_1^e, z_2^e)$ and $F_2(z_3^e, z_4^e)$ be continuous functions such that

$$\begin{aligned}
\dot{z}_1^e &= F_1(z_1^e, z_2^e), \\
\dot{z}_2^e &= -(z_1^e)^3 + z_2^e,
\end{aligned}$$

and

$$\begin{aligned}
\dot{z}_3^e &= F_2(z_3^e, z_4^e), \\
\dot{z}_4^e &= -(z_3^e)^3 + z_4^e,
\end{aligned}$$

respectively, are globally asymptotically stable at the origin. If we show that the system (**3**.4.53), with $u_1 = F_1(z_1^e, z_2^e)$, $u_2 = F_2(z_3^e, z_4^e)$, $u_{11}^w = -z_5^e$ and $u_{21}^w = -z_6^e$, is semiglobally asymptotically stabilizable at the origin through a feedback law $u_{i2}^w = F_i^w(z_7^e)$, $i = 1, 2$, the rest of the proof follows as in the proof of theorem **3**.4.17. For, note that, with $u_{12}^w = 0$, $u_{22}^w = z_7^e - v_{22}^w$ and a global change of coordinates

$$\begin{aligned}
w_i^e &= z_i^e, \qquad i \neq 4, \\
w_4^e &= z_4^e + z_7^e,
\end{aligned}$$

we obtain a system of the form

$$\begin{aligned}
\dot{\zeta}^e &= \varphi(\zeta^e), \\
\dot{w}_7^e &= v_{22}^w,
\end{aligned} \tag{4.22}$$

where $\zeta^e = (\, w_1^e \quad \cdots \quad w_6^e \,)^T$ and $\dot{\zeta}^e = \varphi(\zeta^e)$ is globally asymptotically stable at the origin. Any feedback law $v_{22}^w = -\gamma w_7^e$ with $\gamma > 0$ is such that (4.22) with $v_{22}^w = -\gamma w_7^e$ is globally asymptotically stable at the origin.

4.5 Noninteracting control with stability on compact sets

Sometimes the requirement of global asymptotic stability may be too strong. Yet, one might ask more simply for *local* asymptotic stability at the origin of the state space with domain of attraction containing an *a priori* given *compact* set. For some $n^w \geq 0$ and , We will say that the system (3.1.1) is *semiglobally* asymptotically stabilizable at the origin if there exists $n^w \geq 0$ such that for each compact set $\Omega \subset \mathbb{R}^{n+n^w}$ there exists a feedback law $u = \alpha(x)$ such that the resulting closed–loop system is *locally* asymptotically stable at the origin with domain of attraction containing Ω ([82]). It is well–known that, in general, semiglobal stabilizability does *not* imply global stabilizability.

In analogy with section 3.1, we formulate the problem of achieving noninteraction and stability on compact sets as follows.

Noninteracting control with stability on compact sets via static feedback laws (NSCS). *For any compact set $\Omega \subset \mathbb{R}^n$, find, if possible, a static feedback law (3.1.2), defined for all $x \in \mathbb{R}^n$, such that the system (3.1.1)–(3.1.2) is noninteractive, locally asymptotically stable in x_0, with domain of attraction containing Ω, and has uniform vector relative degree.*

Noninteracting control with stability on compact sets via dynamic feedback laws (NSCD). *Find, if possible, $n^w \geq 0$ such that, for any compact set $\Omega^e \subset \mathbb{R}^{n+n^w}$, there exists a dynamic feedback law (3.1.2), defined for all $(x, w) \in \mathbb{R}^{n+n^w}$, such that the system (3.1.1)–(3.1.2) is noninteractive, locally asymptotically stable in (x_0, w_0), with domain of attraction containing Ω^e, and has uniform vector relative degree.*

We will consider the class of systems which satisfy assumptions (A2), (A3) and (A4) for all $x \in \mathbb{R}^n$, assumption (A5) and admit the global decompositions (3.4.4)–(3.4.5) and (2.4.21). We will not investigate the problem of noninteraction and stability on compact sets via static feedback. The interested reader is referred to remark 5.2 for some possible strategy of solution to this problem.

The property of (4.10) being convergent input convergent state (see theorem 4.5) is somewhat difficult to be checked. However, if the vector fields T_{ik}'s satisfy an additional simple condition and if stability on compact sets is enough for our purposes, we obtain the following interesting result.

Theorem 5.1. Assume that (A2), (A3) and (A4) hold for all $x \in \mathbb{R}^n$. Moreover, assume that (A5) holds and (3.1.1) admits the global decomposition (3.4.4)–(3.4.5). Then, NSCD is solvable if

a) the dynamics (3.4.7) is globally asymptotically stable at the origin,

b) the system (4.1) is convergent input convergent state,

c) each system (3.4.8) is globally asymptotically stabilizable at the origin via dynamic feedback,

d) the vector fields T_{ik} are complete and $[T_{ik}, T_{ih}](x) = 0$ for all x and for $k, h = 1, \ldots, n_0$.

□

Proof. Throughout the proof, we will assume without loss of generality that $\Delta_{MIX} = 0$ (see the proof of theorem 3.4.17).

By assumption d), a feedback law $u^e = \alpha^e(x^e) + \sum_{j=1}^{2m} \beta_j^e(x^e)v_j^e$, which satisfies (3.4.31) and (3.4.32), can be chosen as in the proof of proposition 4.3. By doing this, the vector fields $\widetilde{g}^e_{m+j}$, $j = 1, \ldots, m$ are complete and commute.

Reasoning as in the proof of proposition 4.4, we can find a change of coordinates $z^e = \left((z_1^e)^T(x^e) \quad \cdots \quad (z_{m+1}^e)^T(x)\right)^T$, with $z^e(x_0^e) = 0$ and defined for all $x^e \in \mathbb{R}^{n+n^w}$, such that $\mathcal{R}_i^e = \text{span}\{\frac{\partial}{\partial z_i}\}$ and the system (3.4.51) is in the form (3.4.53). Moreover, from the proof of proposition 4.4 and assumption c), it follows that, without loss of generality, we can assume that each system $\dot{z}_i^e = \widetilde{f}_i^e(z_i^e, 0)$ is globally asymptotically stable at the origin. To prove the theorem, it suffices to prove that the system (3.4.53), with $v_1^e = \ldots = v_m^e = 0$, is semiglobally asymptotically stabilizable at the origin of its state space through a feedback law, which preserves the decoupled structure of the system itself. Since the matrix $\bar{G}(z_{m+1}^e)$ has full row rank for all z_{m+1}^e (see proof of proposition 3.4.16), the system (3.4.53), with $v_1^e = \ldots = v_m^e = 0$ and output vector z_{m+1}^e, is *globally minimum phase* and has uniform vector relative degree one. Since $\widetilde{g}^e_{m+j}$, $j = 1, \ldots, m$, are complete and commute, it follows from corollary 5.6 of [12] that there exists a change of coordinates $\eta^e(z^e) = \left((\eta_1^e)^T(z^e) \quad (z_{m+1}^e)^T\right)^T$, $\eta(0) = 0$, defined for all $x^e \in \mathbb{R}^{n+n^w}$ and such that (3.4.53), with $v_1^e = \ldots = v_m^e = 0$, is expressed in these coordinates by

$$\begin{aligned} \dot{\eta}_1^e &= \bar{f}_1^e(\eta_1^e, z_{m+1}^e), \\ \dot{z}_{m+1}^e &= \widetilde{f}_{m+1}^e(z_{m+1}^e) + \sum_{j=m+1}^{2m} \widetilde{g}^e_{m+1,j}(z_{m+1}^e)v_j^e. \end{aligned} \tag{5.1}$$

From theorem 6.1 of [82] it follows that the system (5.1) is semiglobally asymptotically stabilizable at the origin through a feedback law $v_j^e = F_j^e(z_{m+1}^e)$, $j = m+1, \ldots, 2m$ (with $F_j^e(z_{m+1}^e)$ depending on Ω^e). □

A typical situation in which the vector fields T_{ik} are complete and commute is when the vector fields X_{ik} of the form (3.4.23) are functions only of x_i.

Remark 5.2. In general, if the vector fields T_{ik} either are not complete or do not commute, the proof of theorem 5.1 can be carried through if the system (**3**.4.53), with $v_1^e = \ldots = v_m^e = 0$, is semiglobally asymptotically stabilizable at the origin of its state space through a feedback law which preserves the decoupled structure of the system itself.

The available results on semiglobal stabilization of interconnected systems ([12] and [82]) can be used also to find conditions to solve NSCS. Indeed, NSCS is solvable for (**3**.1.1) if (**3**.3.4) is semiglobally stabilizable at the origin via feedbak laws of the form $u_i = \alpha_i(x_i)$, $i = 1, \ldots, m$ (if the distributions $\langle \tilde{f}, \tilde{g}_1, \ldots, \tilde{g}_m | \text{span}\{\tilde{g}_i\}\rangle$, $i = 1, \ldots, m$, have constant dimension for all $x \in I\!R^n$, also the converse is true). □

Following the proof of the necessary part of theorem 4.5, we can prove also a necessary condition to solve NSCD. Let $d_i = s_i - \dim(\Delta_{MIX}(x) \cap \mathcal{R}_i^*(x))$.

*Theorem 5.3. Assume that (A2), (A3) and (A4) hold for all $x \in I\!R^n$. Moreover, assume that (A5) holds and (**3**.1.1) admits the global decompositions (**3**.4.4)–(**3**.4.5) and (**3**.4.21). Let (**2**.1.2) be a feedback law which solves NSCD. If, in addition, the distributions $\mathcal{R}_i^e = \langle \tilde{f}^e, \tilde{g}_1^e, \ldots, \tilde{g}_m^e | \text{span}\{\tilde{g}_i^e\}\rangle, i = 1, \ldots, m$, are regularly computable at each $x^e \in I\!R^n$ and the distributions Δ_{MIX}^e and $\Delta_{MIX}^e + \mathcal{R}_i^e$, $i = 1, \ldots, m$, have constant dimension for all $x^e \in I\!R^{n+n^w}$, then*

*a) the dynamics (**3**.4.7) is globally asymptotically stable at the origin,*

*b) for each system (**3**.4.8) there exists $k_i \geq 0$ such that, for each compact set $\Omega_i^e \subset I\!R^{d_i+k_i}$, there exists a feedback law*

$$\begin{aligned} u_i &= \eta_i(z_1^i, z_3^i, \xi_i), \\ \dot{\xi}_i &= \zeta_i(z_1^i, z_3^i, \xi_i), \qquad \xi_i \in I\!R^{k_i}, \end{aligned} \tag{5.2}$$

*such that the system (**3**.4.8)–(5.2) is locally asymptotically stable at the origin with domain of attraction containing Ω_i^e.* □

CHAPTER 5

NONINTERACTING CONTROL WITH LOCAL STABILITY FOR NONLINEAR SYSTEMS: THE GENERAL CASE

So far we have been studying the problem of achieving noninteraction with local stability for nonlinear systems with $m = p = \nu$, i.e. each output is controlled by one input. The concept of vector relative degree and the asymptotic properties of some nonlinear dynamics have been playing a key role in solving this problem, either near the origin of the state space or in the large.

One might ask what kind of results can be established for larger classes of nonlinear systems (for example, when either $m \neq p$ or $p \neq \nu$). Few basic question must be answered, first. Is the class of invertible feedback laws still a *good* one to obtain significant results? Does the definition of vector relative degree still make sense and, if not, which is its natural substitute in more general situations? Are the nonlinear dynamics (**3**.4.7) still feedback invariant and, if not, which is its natural extension?

A first step in this direction is to consider nonlinear systems

$$\begin{aligned} \dot{x} &= f(x) + \sum_{j=1}^{m} g_j(x) u_j, \\ y_i &= h_i(x), \qquad i = 1, \ldots, p, \end{aligned} \tag{1.1}$$

with $p = \nu$, i.e. each output is controlled by possibly more than one input. We will see in the following two sections that, with the same formulation as the one given in section **3**.2, most of the results of chapter **2** can be extended to this class of systems. The last section is dedicated to the most general case $p > \nu$ (a new formulation of the problem will be required in this case).

In the next few lines, we give some basic notations, which will be freely used throughout this chapter.

Let $G(x)$, $\mathcal{G}$ and $\mathcal{K}_i$ as in chapter **3**. Given a feedback law (**3**.1.2) and an integer s such that $m \geq s \geq 1$, let us partition $\alpha(x,w)$, $\gamma(x,w)$, $\beta(x,w)$ and $\delta(x,w)$ into blocks as follows

$$\alpha(x,w) = (\, \alpha_1^T(x,w) \quad \cdots \quad \alpha_s^T(x,w)\,)^T \ , \ \beta(x,w) = (\, \beta_1(x,w) \cdots \ \ \beta_s(x,w)\,),$$

$$\gamma(x,w) = (\gamma_1^T(x,w) \quad \cdots \quad \gamma_s^T(x,w))^T \ , \ \delta(x,w) = (\delta_1(x,w) \cdots \quad \delta_s(x,w)) \, ;$$

Moreover, let

$$\widetilde{f}^e(x,w) = \begin{pmatrix} f(x) + g(x)\alpha(x,w) \\ \gamma(x,w) \end{pmatrix},$$
$$\widetilde{G}^e(x,w) = \begin{pmatrix} G\beta(x,w) \\ \delta(x,w) \end{pmatrix}$$

and, according to the above partition,

$$\widetilde{g}_j^e(x,w) = \begin{pmatrix} G(x)\beta_j(x,w) \\ \delta_j(x,w) \end{pmatrix}, \qquad j = 1, \dots, s.$$

In the case $n^w = 0$, we obtain static feedback laws and, more simply, we replace $\widetilde{f}^e(x,w)$ by $\widetilde{f}(x)$, $\widetilde{g}_j^e(x,w)$ by $\widetilde{g}_j(x)$ and $\widetilde{G}^e(x,w)$ by $\widetilde{G}(x)$.

Moreover, with abuse of notation, if $X_1, \dots, X_k$ and $Y_1, \dots, Y_l$ are smooth vector fields of $U \subset I\!R^n$ and $X(x) = (X_1(x) \quad \cdots X_k(x))$ and $Y(x) = (Y_1(x) \quad \cdots \quad Y_l(x))$, we denote by $[X,Y](x)$ the matrix with $[X_i, Y_j](x)$ as $[2(i-1)+j]$–th column.

5.1 Noninteracting control with local stability via static feedback

In this section we will give some necessary and sufficient conditions to solve NLSS, as formulated in section 3.2, for the class of systems (1.1) with $p = \nu$. Moreover, we will consider the class of regular feedback laws $u = \alpha(x) + \beta(x)v$. Although there is no practical reason for so constraining the class of admissible feedback laws, this assumption allows us to extend some results of section 3.3 to the class of systems here considered.

Throughout the chapter, we will assume that (A1), (A2) and (A3) hold. (A1) is a necessary and sufficient condition to solve NS even in the case $p = \nu$ (see [55], remark 6.3.15). Moreover, we assume that

(A2b) $\mathcal{Q}^*$, the maximal controllability contained in $\mathcal{R}^*$, is regularly computable at x_0.

5.1.1 Regular solutions and invariant dynamics induced by a regular solution

We have seen that, when $m = p = \nu$, the set of controllability distributions $\mathcal{R}_1^*, \dots, \mathcal{R}_\nu^*$ are always rendered invariant by a feedback law which solves NS (proposition 3.3.1). In general, when $m \neq p$ or $p \neq \nu$, this is *not* true. To see this, let us consider the following linear system

$$\begin{aligned} \dot{x}_1 &= -x_4 + u_1, \\ \dot{x}_2 &= u_2, \\ \dot{x}_3 &= x_1 + x_2 + x_3, \\ \dot{x}_4 &= x_5, \\ \dot{x}_5 &= u_3, \\ y_i &= x_i, \qquad i = 1, 2. \end{aligned} \tag{1.2}$$

Clearly, (1.2) is noninteractive, with $I_1 = \{1,3\}$ and $I_2 = \{2\}$. Moreover,

$$\mathcal{R}_1^* = \text{span}\{\begin{pmatrix} 1 & 0 & 0 & 0 \\ 0 & 0 & 0 & 0 \\ 0 & 1 & 0 & 0 \\ 0 & 0 & 1 & 0 \\ 0 & 0 & 0 & 1 \end{pmatrix}\}, \ \mathcal{R}_2^* = \text{span}\{\begin{pmatrix} 0 & 0 & 0 & 0 \\ 1 & 0 & 0 & 0 \\ 0 & 1 & 0 & 0 \\ 0 & 0 & 1 & 0 \\ 0 & 0 & 0 & 1 \end{pmatrix}\}, \ \mathcal{Q}^* = \text{span}\{\begin{pmatrix} 0 & 0 \\ 0 & 0 \\ 0 & 0 \\ 1 & 0 \\ 0 & 1 \end{pmatrix}\}.$$

Note that $\mathcal{R}_2^*$ is *not* invariant. Nonetheless, the controllability subspaces

$$\mathcal{R}_1 = \mathcal{R}_1^*, \ \mathcal{R}_2 = \text{span}\{\begin{pmatrix} 0 & 0 \\ 1 & 0 \\ 0 & 1 \\ 0 & 0 \\ 0 & 0 \end{pmatrix}\},$$

are invariant and satisfy $\mathcal{R}_i \subset \bigcap_{j \neq i} \mathcal{K}_j$, $i = 1, 2$. Note that the feedback law $u_1 = x_4 + v_1$ renders invariant the distributions $\mathcal{R}_1^*, \mathcal{R}_2^*$ and does not destroy the noninteraction property of (1.2).

The above example suggests that a fundamental role in solving NS is played by the family of compatible controllability distributions. Indeed, it is not hard to show that, as in the case of linear systems, the problem of achieving noninteraction amounts to finding ν controllability distributions $\mathcal{R}_1, \ldots, \mathcal{R}_\nu$ which are *compatible* and satisfy $\mathcal{R}_i \subset \bigcap_{j \neq i} \mathcal{K}_j$, $i = 1, \ldots, \nu$ (for, combine theorem **3**.3.5 and lemma **1**.8.8 of [55]). Let us state this formally.

Proposition 1.1. Let us consider the class of regular feedback laws. If there exist ν controllability distributions $\mathcal{R}_1, \ldots, \mathcal{R}_\nu$, a static feedback law $u = \alpha(x) + \beta(x)v$ and a block partition $\beta_1(x), \ldots, \beta_{\nu+1}(x)$ of $\beta(x)$ such that

$$\mathcal{R}_i \subset \bigcap_{j \neq i} \mathcal{K}_j, \qquad i = 1, \ldots, \nu, \tag{1.3}$$

$$\mathcal{R}_i = \langle \tilde{f}, \tilde{g}_1, \ldots, \tilde{g}_{\nu+1} | \text{span}\{\tilde{g}_i, \tilde{g}_{\nu+1}\} \rangle, \tag{1.4}$$

then (1.1) can be rendered noninteractive via static feedback. Conversely, assume that (1.1) can be rendered noninteractive via static feedback. If, in addition, the distributions $\langle \tilde{f}, \tilde{g}_1, \ldots, \tilde{g}_{\nu+1} | \text{span}\{\tilde{g}_i, \tilde{g}_{\nu+1}\}\rangle$, $i = 1, \ldots, \nu$, *have constant dimension for all* x *in a neighbourhood of* x_0, *then there exist* ν *controllability distributions* $\mathcal{R}_1, \ldots, \mathcal{R}_\nu$, *a static feedback law* $u = \alpha(x) + \beta(x)v$ *and a block partition* $\beta_1(x), \ldots, \beta_{\nu+1}(x)$ *of* $\beta(x)$ *which satisfy (1.3) and (1.4).* □

The following proposition gives a simple necessary and sufficient conditions for the existence of a set of controllability distributions, which are compatible and satisfy (1.3).

Proposition 1.2. Assume that (A2), (A2b) and (A3) hold and consider the class of regular feedback laws. If there exist ν *controllability distributions* $\mathcal{R}_1, \ldots, \mathcal{R}_\nu$ *such that*

$$\mathcal{R}_i \subset \bigcap_{j \neq i} \mathcal{K}_j,$$

$$\mathcal{G} = \sum_{j=1}^{\nu} (\mathcal{R}_j \cap \mathcal{G}), \tag{1.5}$$

then there exist a static feedback law $u = \alpha(x) + \beta(x)v$ *and a block partition* $\beta_1(x), \ldots, \beta_{\nu+1}(x)$ *of* $\beta(x)$ *which satisfy (1.3) and (1.4).*

Conversely, suppose that there exist ν *controllability distributions* $\mathcal{R}_1, \ldots, \mathcal{R}_\nu$, *a feedback law* $u = \alpha(x) + \beta(x)v$ *and a block partition* $\beta_1(x), \ldots, \beta_{\nu+1}(x)$ *of* $\beta(x)$ *which satisfy (1.3) and (1.4). If, in addition, the distributions,* $\langle \tilde{f}, \tilde{g}_1, \ldots, \tilde{g}_{\nu+1} | \text{span}\{\tilde{g}_i, \tilde{g}_{\nu+1}\}\rangle$, $i = 1, \ldots, \nu$, *have constant dimension in a neighbourhood of* x_0, *then the distributions* $\mathcal{R}_1, \ldots, \mathcal{R}_\nu$ *satisfy (1.3) and (1.5).* □

Proof. The necessity is a trivial consequence of theorem 3.3.5 and lemma 1.8.8 of [55]. The sufficiency can be proved as in [62]. □

It is well–know that, under assumptions (A2), (A2b) and (A3), conditions (1.4) and (1.5) are equivalent to (A1) (see for example [26]).

The existence of ν controllability distributions $\mathcal{R}_1, \ldots, \mathcal{R}_\nu$, which are compatible and satisfy (1.3), implies (and is implied by the fact) that also the distributions $\mathcal{R}_1^*, \ldots, \mathcal{R}_\nu^*$ are compatible (use propositions 1.1 and 1.2 and the fact that $\mathcal{R}_i \subset \mathcal{R}_i^*$, $i = 1, \ldots, \nu$). However, as shown by the example above, the distributions $\mathcal{R}_1^*, \ldots, \mathcal{R}_\nu^*$ may be *not* rendered invariant by a feedback law which solves NS.

Following [40], we will say that a set of controllability distributions $\mathcal{R}_1, \ldots, \mathcal{R}_\nu$ is a *regular set* (at x_0) if it satisfies assumptions (A2) and (A2b) (with $\mathcal{R}_1^*, \ldots, \mathcal{R}_\nu^*$ and $\mathcal{Q}^*$

replaced by $\mathcal{R}_1,\dots,\mathcal{R}_\nu$ and $\mathcal{Q}$, respectively). Moreover, a set of controllability distributions $\mathcal{R}_1,\dots,\mathcal{R}_\nu$ is said to be a *regular solution* (of NS at x_0) if it is a regular set and, in addition, it satisfies (1.3) and (1.5). Note that, since $\mathcal{R}_i \subset \mathcal{R}_i^*$ and from (A2) and (A2b), it follows that, if $\mathcal{R}_1,\dots,\mathcal{R}_\nu$ is a regular solution, also $\mathcal{R}_1^*,\dots,\mathcal{R}_\nu^*$ is.

At this point, one might ask how *thick* is the set of regular solutions. It is not hard to show that, when $m = p = \nu$, if the regular set $\mathcal{R}_1^*,\dots,\mathcal{R}_\nu^*$ is a regular solution, then it is also the *only one* with this property (see corollary 1.4). This is *not* true in general and the next proposition describes how any regular solution relates to the maximal regular solution $\mathcal{R}_1^*,\dots,\mathcal{R}_\nu^*$.

Proposition 1.3. Assume that $\mathcal{R}_1,\dots,\mathcal{R}_\nu$ and $\mathcal{R}_1^*,\dots,\mathcal{R}_\nu^*$ are regular solutions. Moreover, assume that the distributions $\mathcal{R}_i+\mathcal{Q}^*$, $((\mathcal{R}_i+\mathcal{Q}^*)\cap\mathcal{G})\cap(\sum_{j\neq i}((\mathcal{R}_j+\mathcal{Q}^*)\cap\mathcal{G}))$, $(\mathcal{R}_i^*\cap\mathcal{G})\cap(\sum_{j\neq i}((\mathcal{R}_j+\mathcal{Q}^*)\cap\mathcal{G}))$ and $\sum_{j\neq i}((\mathcal{R}_j+\mathcal{Q}^*)\cap\mathcal{G})$, $i=1,\dots,\nu$, have constant dimension in a neighbourhood of x_0. Then,

$$\mathcal{R}_i+\mathcal{Q}^* = \mathcal{R}_i^*, \qquad i=1,\dots,\nu. \qquad \square$$

Proof. The proof follows the same lines of the proof of theorem 2.6 of [40] with some nontrivial changes. First, note that $\mathcal{Q}^* \subset \mathcal{R}_i^*$ for $i=1,\dots,\nu$. Indeed, $\mathcal{Q}^*$ is, by definition, a controllability distribution contained in $\bigcap_{j=1}^{\nu}\mathcal{K}_j$, which, in turn, is contained in $\bigcap_{j\neq i}\mathcal{K}_j$ for $i=1,\dots,\nu$. Our claim follows from the fact that $\mathcal{R}_i^*$ is the maximal controllability distribution contained in $\bigcap_{j\neq i}\mathcal{K}_j$. It is not hard to show that

$$(\mathcal{R}_i+\mathcal{Q}^*)\cap\mathcal{G} = \mathcal{R}_i^*\cap\mathcal{G}, \qquad i=1,\dots,\nu. \tag{1.6}$$

For, note that

$$\mathcal{Q}^*\cap\mathcal{G} \subset ((\mathcal{R}_i+\mathcal{Q}^*)\cap\mathcal{G})\cap(\sum_{j\neq i}((\mathcal{R}_j+\mathcal{Q}^*)\cap\mathcal{G})) \subset (\mathcal{R}_i^*\cap\mathcal{G})\cap(\sum_{j\neq i}((\mathcal{R}_j+\mathcal{Q}^*)\cap\mathcal{G})) \subset$$
$$\subset (\mathcal{R}_i^*\cap\mathcal{G})\cap(\sum_{j\neq i}\mathcal{R}_j^*)\cap\mathcal{G} \subset \mathcal{R}^*\cap\mathcal{G} = \mathcal{Q}^*\cap\mathcal{G},$$

which implies that

$$((\mathcal{R}_i+\mathcal{Q}^*)\cap\mathcal{G})\cap(\sum_{j\neq i}((\mathcal{R}_j+\mathcal{Q}^*)\cap\mathcal{G})) = (\mathcal{R}_i^*\cap\mathcal{G})\cap(\sum_{j\neq i}((\mathcal{R}_j+\mathcal{Q}^*)\cap\mathcal{G})).$$

From the regularity of $\mathcal{R}_1^*,\dots,\mathcal{R}_\nu^*$ and $\mathcal{R}_1,\dots,\mathcal{R}_\nu$ we obtain

$$\mathcal{R}_i^*\cap\mathcal{G} + \sum_{j\neq i}((\mathcal{R}_j+\mathcal{Q}^*)\cap\mathcal{G}) = \mathcal{G} = (\mathcal{R}_i+\mathcal{Q}^*)\cap\mathcal{G} + \sum_{j\neq i}((\mathcal{R}_j+\mathcal{Q}^*)\cap\mathcal{G}).$$

By an easy dimensionality argument we obtain (1.6). From here we can proceed exactly as in theorem **2**.6 of [40]. □

It is not surprising that, when $m = p = \nu$, $\mathcal{Q}^* = 0$. As a consequence of proposition 1.3, we conclude that the maximal regular solution $\mathcal{R}_1^*, \ldots, \mathcal{R}_\nu^*$ is the *only* regular solution. This is shown in the following corollary of proposition 1.3 (a more complex proof has been given in [62]).

Corollary 1.4. Assume $m = p = \nu$ and let (A1), (A2), (A2b) and (A3) hold. Then, $\mathcal{Q}^* = 0$ and $\mathcal{R}_1^*, \ldots, \mathcal{R}_\nu^*$ is the only regular solution. □

Proof. Note that $\mathcal{Q}^* \cap \mathcal{G} = \mathcal{R}^* \cap \mathcal{G}$. Since (1.1) has vector relative degree $r_1, \ldots, r_m$ at x_0, it follows that $\mathcal{V}^* = (\text{span}\{dh_i, \ldots, dL_f^{r_i-1}h_i : i = 1, \ldots, m\})^\perp$, where $\mathcal{V}^*$ denotes the maximal controlled invariant distribution contained in $\bigcap_{i=1}^{m}\mathcal{K}_i$ ([55], proposition **2**.3.1). Since the distributions $\mathcal{R}_1^*, \ldots, \mathcal{R}_\nu^*$ are compatible, $\mathcal{R}^*$ is controlled invariant and is contained in $\bigcap_{i=1}^{m}\mathcal{K}_i$. Thus, by definition $\mathcal{R}^* \subset \mathcal{V}^*$. Since the decoupling matrix of (1.1) is nonsingular at x_0, it follows that $\mathcal{V}^* \cap \mathcal{G} = 0$, which proves $\mathcal{Q}^* = 0$. □

Now, suppose that $\mathcal{R}_1, \ldots, \mathcal{R}_\nu$ be a regular solution. By proposition 1.2 it follows that there exist a feedback law $u = \alpha(x) + \beta(x)v$ and a block partition $\beta_1(x), \ldots, \beta_{\nu+1}(x)$ of $\beta(x)$ such that

$$\mathcal{R}_i = \text{span}\{\widetilde{f}, \widetilde{g}_1, \ldots, \widetilde{g}_{\nu+1} | \text{span}\{\widetilde{g}_i, \widetilde{g}_{\nu+1}\}\rangle, \qquad i = 1, \ldots, \nu. \tag{1.7}$$

In analogy with proposition (**3**.3.2), we can prove the following result, which completely characterizes the family of regular feedback laws satisfying (1.7).

Proposition 1.5. Assume that (A1), (A2), (A2b) and (A3) hold and let $\mathcal{R}_1, \ldots, \mathcal{R}_\nu$ be a regular solution. Moreover, let $u = \alpha^h(x) + \beta^h(x)v$, $h = 1, 2$, with

$$\alpha^h(x) = \left((\alpha_1^h)^T(x) \quad \cdots \quad (\alpha_{\nu+1}^h)^T(x) \right)^T,$$
$$\beta^h(x) = \left(\beta_1^h(x) \quad \cdots \quad \beta_{\nu+1}^h(x) \right),$$

be two regular feedback laws which satisfy (1.7). Then, $\alpha^2(x) - \alpha^1(x) = \sum_{j=1}^{\nu+1} \beta_j^1(x)\alpha_j^0(x)$, where each entry of $\alpha_i^0(x)$ is a smooth real-valued function, constant along each leaf of $\sum_{j\neq i}\mathcal{R}_j$ and equal to zero for $x \in \mathcal{L}_{x_0}^{\sum_{j\neq i}\mathcal{R}_j}$, and $\beta^2(x) = \beta^1(x)\beta^0(x)$, where $\beta^0(x)$ is a block diagonal matrix such that each diagonal block is nonsingular at x_0 and each entry of the i-th diagonal block of $\beta^0(x)$ is a smooth real-valued function, constant along each leaf of $\sum_{j\neq i}\mathcal{R}_j$. □

Let $\mathcal{R}_1, \ldots, \mathcal{R}_\nu$ be a regular solution with $\mathcal{R} = \bigcap_{i=1}^{\nu} \sum_{j \neq i} \mathcal{R}_j$ and $\mathcal{Q}$ be the maximal controllability distribution contained in $\mathcal{R}$. Denote by $\widetilde{\Sigma}$ the system, resulting from plugging into (1.1) a feedback law which satisfies (1.7). Following section **3**.3, it is possible to express $\widetilde{\Sigma}$ in a *standard* noninteractive form.

Proposition 1.6. Assume that (A1), (A2), (A2b) and (A3) hold. Then, there exists a change of coordinates $z(x) = (\, z_1^T(x) \quad \cdots \quad z_{\nu+1,1}^T(x) \quad z_{\nu+1,2}^T(x)\,)^T$, $z(x_0) = 0$, *defined in a neighbourhood of* x_0 *and such that* $\widetilde{\Sigma}$ *is expressed in these coordinates by*

$$\begin{aligned} \dot{z}_i &= \widetilde{f}(z_i) + \widetilde{g}_{ii}(z_i)v_i, \qquad i = 1, \ldots, m, \\ \dot{z}_{\nu+1,1} &= \widetilde{f}_{\nu+1,1}(z_1, \ldots, z_{\nu+1,1}) + \sum_{j=1}^{\nu} \widetilde{g}_{\nu+1,1j}(z_1, \ldots, z_{\nu+1,1})v_j, \\ \dot{z}_{\nu+1,2} &= \widetilde{f}_{\nu+1,2}(z) + \sum_{j=1}^{\nu+1} \widetilde{g}_{\nu+1,2j}(z)v_j, \\ y_i &= h_i(z_i), \qquad i = 1, \ldots, \nu, \end{aligned} \tag{1.8}$$

with $\sum_{j \neq i} \mathcal{R}_j = \operatorname{span}\{\partial/\partial z_j : j \neq i\}$, $i = 1, \ldots, \nu$, $\mathcal{R} = \operatorname{span}\{\partial/\partial z_{\nu+1,1}, \partial/\partial z_{\nu+1,2}\}$ *and* $\mathcal{Q} = \operatorname{span}\{\partial/\partial z_{\nu+1,2}\}$. □

In what follows, for simplicity of notation, we let $z = x$ and $v = u$. In analogy with section **3**.3, we introduce the following nonlinear dynamics. Let us consider the restriction $\widetilde{f}|\mathcal{L}_{x_0}^{\mathcal{R}}$. This vector field induces a dynamics evolving on $\mathcal{L}_{x_0}^{\mathcal{R}}$, which in x–coordinates is described by

$$\begin{aligned} \dot{x}_{\nu+1,1} &= \widetilde{f}_{\nu+1,1}(0, \ldots, 0, x_{\nu+1,1}), \\ \dot{x}_{\nu+1,2} &= \widetilde{f}_{\nu+1,2}(0, \ldots, 0, x_{\nu+1,1}, x_{\nu+1,2}), \end{aligned}$$

When $\mathcal{R}_i = \mathcal{R}_i^*$, $i = 1, \ldots, m$, the dynamics

$$\dot{x}_{\nu+1,1} = \widetilde{f}_{\nu+1,1}(0, \ldots, 0, x_{\nu+1,1}) \tag{1.9}$$

and

$$\dot{x}_i = \widetilde{f}(x_i) + \widetilde{g}_{ii}(x_i)u_i, \tag{1.10}$$

are the nonlinear analogue of (**2**.2.3) and (**2**.2.4), respectively. Moreover, it makes sense to consider the restrictions $\widetilde{f}|\mathcal{L}_{x_0}^{\mathcal{Q}}$ and $\widetilde{g}_{\nu+1}|\mathcal{L}_{x_0}^{\mathcal{Q}}$ (the latter since $\widetilde{g}_{\nu+1} \in \mathcal{Q}$). These vector fields induce a dynamics on $\mathcal{L}_{x_0}^{\mathcal{Q}}$, which in x–coordinates (1.8) is described by

$$\dot{x}_{\nu+1,2} = \widetilde{f}_{\nu+1,2}(0, \ldots, 0, x_{\nu+1,2}) + \widetilde{g}_{\nu+1,2,\nu+1}(0, \ldots, 0, x_{\nu+1,2})u_{\nu+1}. \tag{1.11}$$

When $\mathcal{R}_i = \mathcal{R}_i^*$, the dynamics (1.11) is, in the case of linear systems, exactly (2.2.5).

The following proposition states that the dynamics (1.9), (1.10) and (1.11) are *uniquely* (in some sense) associated to a given system (1.1), which satisfies (A1), (A2), (A2b) and (A3), and to a given regular solution (the proof is similar to that of proposition 3.3.4). Given a regular solution $\mathcal{R}_1, \ldots, \mathcal{R}_\nu$ and the corresponding system (1.8), let us apply to (1.8) a regular feedback law $u = \alpha'(x) + \beta'(x)v$, which satisfies (1.7). Denote by $\tilde{\Sigma}'$ the system obtained from $\tilde{\Sigma}$ and by (1.9)', (1.10)' and (1.11)' the dynamics corresponding to (1.9), (1.10) and (1.11), respectively, defined on $\tilde{\Sigma}'$.

Proposition 1.7. The dynamics (1.9)–(1.9)' are equal up to changes of coordinates, defined in a neighbourhood of x_0. The systems (1.10)–(1.10)' and (1.11)–(1.11)', respectively, are equal up to changes of coordinates and invertible static feedback transformations. □

While proposition 1.7 essentially restates in a more general context what we already know in the case $m = p = \nu$, however we are left wondering if there is any significant relation among the dynamics (1.9) (or (1.10)), respectively associated to *different* regular solutions. We will show that the dynamics (1.9), associated to the maximal regular solution $\mathcal{R}_1^*, \ldots, \mathcal{R}_\nu^*$, is a *subdynamics* of the corresponding dynamics (1.9), associated to any other regular solution $\mathcal{R}_1, \ldots, \mathcal{R}_\nu$. On the other hand, we will show that the dynamics (1.10), associated to a given regular solution $\mathcal{R}_1, \ldots, \mathcal{R}_\nu$, is *equal* (in some sense) to the corresponding dynamics (1.10), associated to the maximal regular solution $\mathcal{R}_1^*, \ldots, \mathcal{R}_\nu^*$.

Toward this end, we need the following result.

Proposition 1.8. Assume that $\mathcal{R}_1, \ldots, \mathcal{R}_\nu$ and $\mathcal{R}_1^*, \ldots, \mathcal{R}_\nu^*$ be regular solutions. Moreover, assume that the distribution $\mathcal{R} \cap \mathcal{Q}^*$ has constant dimension in a neighbourhood of x_0. Then,

$$\mathcal{R} + \mathcal{Q}^* = \mathcal{R}^*.$$ □

Proof. The proof follows the same lines of the proof of theorem 3.7 of [40] with some nontrivial changes. Throughout the proof, given a set of distributions $\Delta_1, \ldots, \Delta_r$, linearly independent at each x of a neighbourhood of x_0, we will denote by $\Delta_1 \oplus \ldots \oplus \Delta_r$ the distribution which assigns to each x the subspace $\Delta_1(x) \oplus \ldots \oplus \Delta_r(x)$ or, equivalently, $\bigoplus_{i=1}^{r}$. Suppose for a moment that there exist smooth distributions $\mathcal{G}_1^1, \mathcal{G}_1^{(2)}, \ldots, \mathcal{G}_\nu^1, \mathcal{G}_\nu^2, \mathcal{G}_{\nu+1}$

such that

$$
\begin{aligned}
\mathcal{G} &= \mathcal{G}_1^1 \oplus \ldots \oplus \mathcal{G}_\nu^1 \oplus \mathcal{G}_1^2 \oplus \ldots \oplus \mathcal{G}_\nu^2 \oplus \mathcal{G}_{\nu+1}, \\
\mathcal{R}_i \cap \mathcal{G} &= \mathcal{G}_i^1 \oplus \mathcal{G}_i^2 \oplus \mathcal{G}_{\nu+1}, \qquad i = 1, \ldots, \nu, \\
\mathcal{R}_i^* \cap \mathcal{G} &= \mathcal{G}_i^1 \oplus \mathcal{G}_i^2 \bigoplus_{j \neq i} \mathcal{G}_j^2 \oplus \mathcal{G}_{\nu+1}, \qquad i = 1, \ldots, \nu, \\
\mathcal{Q} \cap \mathcal{G} &= \mathcal{G}_{\nu+1}, \\
\mathcal{Q}^* \cap \mathcal{G} &= \mathcal{G}_1^2 \oplus \ldots \oplus \mathcal{G}_\nu^2 \oplus \mathcal{G}_{\nu+1}.
\end{aligned}
\tag{1.12}
$$

We claim that, with the above decomposition, there exist a feedback law $u = \alpha(x) + \beta(x)v$, regular at x_0, and a block partition $\beta_1^1(x), \beta_1^2(x), \ldots, \beta_\nu^1(x), \beta_\nu^2(x), \beta_{\nu+1}(x)$ of $\beta(x)$ such that

$$
\begin{aligned}
[f + G\alpha, \mathcal{R}_i] &\subset \mathcal{R}_i, \qquad i = 1, \ldots, \nu, \\
[G\beta_j^h, \mathcal{R}_i] &\subset \mathcal{R}_i, \qquad j, i = 1, \ldots, \nu + 1,\ h = 1, 2, \\
[f + G\alpha, \mathcal{R}_i^*] &\subset \mathcal{R}_i^*, \qquad i = 1, \ldots, \nu, \\
[G\beta_j^h, \mathcal{R}_i^*] &\subset \mathcal{R}_i^*, \qquad j, i = 1, \ldots, \nu + 1,\ h = 1, 2,
\end{aligned}
\tag{1.13a}
$$

$$
\begin{aligned}
\mathcal{G}_j^h &= \text{span}\{G\beta_j^h\}, \qquad h = 1, 2, \,, j = 1, \ldots, \nu, \\
\mathcal{G}_{\nu+1} &= \text{span}\{G\beta_{\nu+1}\}.
\end{aligned}
\tag{1.13b}
$$

For, since $\mathcal{R}_1, \ldots, \mathcal{R}_\nu$ and $\mathcal{R}_1^*, \ldots, \mathcal{R}_\nu^*$ are regular solutions, they are also compatible (proposition 1.2). Reasoning as in proposition **2**.3.1, it follows that the distributions $\sum_{j \neq i} \mathcal{R}_j$ and $\sum_{j \neq i} \mathcal{R}_j^*$, $i = 1, \ldots, \nu$, are involutive and controlled invariant. Let $\mathcal{G}_i^1 + \mathcal{G}_i^2 = \mathcal{G}_i$. From (1.12) it follows, as in the proof of lemma 3.5 of [40], that there exist a regular feedback law $u = \alpha(x) + \beta(x)v$ and a block partition $\beta_1(x), \ldots, \beta_{\nu+1}(x)$ of $\beta(x)$ such that

$$
\begin{aligned}
[f + G\alpha, \sum_{j \neq i} \mathcal{R}_j] &\subset \sum_{j \neq i} \mathcal{R}_j, \qquad i = 1, \ldots, \nu, \\
[G\beta_h, \sum_{j \neq i} \mathcal{R}_j] &\subset \sum_{j \neq i} \mathcal{R}_j, \qquad h = 1, \ldots, \nu + 1,\ i = 1, \ldots, \nu, \\
[f + G\alpha, \sum_{j \neq i} \mathcal{R}_j^*] &\subset \sum_{j \neq i} \mathcal{R}_j^*, \qquad i = 1, \ldots, \nu, \\
[G\beta_h, \sum_{j \neq i} \mathcal{R}_j^*] &\subset \sum_{j \neq i} \mathcal{R}_j^*, \qquad h = 1, \ldots, \nu + 1,\ i = 1, \ldots, \nu, \\
\mathcal{G}_i &= \text{span}\{G\beta_i\}, \qquad i = 1, \ldots, \nu + 1.
\end{aligned}
$$

Reasoning as in lemma 3.10 of [62], we conclude that

$$\begin{aligned}
[f+G\alpha,\mathcal{R}_i] &\subset \mathcal{R}_i, & i&=1,\dots,\nu,\\
[G\beta_j,\mathcal{R}_i] &\subset \mathcal{R}_i, & j&=1,\dots,\nu+1\,,\ i=1,\dots,\nu,\\
[f+G\alpha,\mathcal{R}_i^*] &\subset \mathcal{R}_i^*, & i&=1,\dots,\nu,\\
[G\beta_j,\mathcal{R}_i^*] &\subset \mathcal{R}_i^*, & j&=1,\dots,\nu+1\,,\ i=1,\dots,\nu,\\
\mathcal{G}_i &= \mathrm{span}\{G\beta_i\}, & i&=1,\dots,\nu+1.
\end{aligned} \tag{1.14}$$

Now, let us denote by $\widetilde{\Sigma}$ the system (1.1) with $u=\alpha(x)+\beta(x)v$. Moreover, let $v_1,\dots,v_{\nu+2}$ be the block partition of the input vector v corresponding to the above partition of $\beta(x)$. Reasoning as in proposition **3**.3.3 and from (1.14), it easily follows that there exists a change of coordinates $\theta(x)=(\,\theta_{11}^T(x)\quad\theta_{12}^T(x)\quad\cdots\quad\theta_{\nu1}^T(x)\quad\theta_{\nu2}^T(x)\quad\theta_{\nu+1}^T(x)\,)^T$, $\theta(x_0)=0$, defined in a neighbourhood of x_0 and such that $\widetilde{\Sigma}$ is expressed in these coordinates by

$$\begin{aligned}
\dot\theta_{i1} &= \bar f_{i1}(\theta_{i1})+\bar g_{i1}(\theta_{i1})u_i, \qquad i=1,\dots,\nu,\\
\dot\theta_{i2} &= \bar f_{i2}(\theta_{i1},\theta_{i2})+\bar g_{i2}(\theta_{i1},\theta_{i2})u_i, \qquad i=1,\dots,\nu,\\
\dot\theta_{\nu+1} &= \bar f_{\nu+1}(\theta)+\sum_{j=1}^{\nu+1}\bar g_{\nu+1,j}(\theta)u_j,
\end{aligned} \tag{1.15}$$

with $\sum_{j\neq i}\mathcal{R}_j=\mathrm{span}\{\partial/\partial\theta_{jh},\partial/\partial\theta_{\nu+1} : h=1,2\,,\ j\neq i\}$, $\sum_{j\neq i}\mathcal{R}_j^*=\mathrm{span}\{\partial/\partial\theta_{jh},\partial/\partial\theta_{i2},\partial/\partial\theta_{\nu+1} : h=1,2\,,\ j\neq i\}$, $\mathcal{R}=\mathrm{span}\{\partial/\partial\theta_{\nu+1}\}$ and $\mathcal{R}^*=\mathrm{span}\{\partial/\partial\theta_{\nu+1},\partial/\partial\theta_{j2} : j=1,\dots,\nu\}$. By direct inspection of (1.15), since $\mathcal{G}_i^2\subset\mathcal{Q}^*\cap\mathcal{G}$ and $\bar g_{i1}(\theta_{i1})$ is a function only of θ_{i1}, it follows that the matrices $\beta_j(x)$, $j=1,\dots,\nu+1$, can be chosen in such a way to satisfy (1.13). This proves our claim.

Now, assume that $\mathcal{Q}^*\cap\mathcal{R}=0$ (in the case $\mathcal{Q}^*\cap\mathcal{R}\neq 0$, we can proceed exactly as in section **A**.3.b of [40]) and denote $f+G\alpha$ by $\widetilde f$, $G\beta_j^i$ by $\widetilde g_j^i$ and $G\beta_{\nu+1}$ by $\widetilde g_{\nu+1}$. Moreover, let $\mathcal{Q}^*=\mathcal{S}_1\oplus\ldots\oplus\mathcal{S}_{\nu+1}$, where $\mathcal{S}_i\subset\langle\widetilde f,\widetilde g_1^1,\widetilde g_1^2,\dots,\widetilde g_\nu^1,\widetilde g_\nu^2,\widetilde g_{\nu+1}|\mathrm{span}\{\widetilde g_i^2\}\rangle$ and $\mathcal{S}_{\nu+1}\subset\langle\widetilde f,\widetilde g_1^1,\widetilde g_1^2,\dots,\widetilde g_\nu^1,\widetilde g_\nu^2,\widetilde g_{\nu+1}|\mathrm{span}\{\widetilde g_{\nu+1}\}\rangle$.

Since $\mathcal{R}+\mathcal{Q}^*\subset\mathcal{R}^*$, it suffices to show that $\dim(\mathcal{R}^*(x))\leq\dim(\mathcal{R}(x))+\dim(\mathcal{Q}^*(x))$. By proposition 1.3, we have $\sum_{j\neq i}\mathcal{R}_j^*=\sum_{j\neq i}\mathcal{R}_j+\mathcal{Q}^*$ or, equivalently, $\sum_{j\neq i}\mathcal{R}_j^*=\sum_{j\neq i}\mathcal{R}_j+\mathcal{S}_i$. Moreover, from lemma **A**.3 of [38] and since $\mathcal{R}_1,\dots,\mathcal{R}_\nu$ is a regular solution, $\dim(\mathcal{R}(x))=\sum_{i=1}^{\nu}\dim(\sum_{j\neq i}\mathcal{R}_j(x))-(\nu-1)n$ (in particular, this identity holds with $\mathcal{R}_i=\mathcal{R}_i^*$ and $\mathcal{R}=\mathcal{R}^*$). Since $\mathcal{S}_{\nu+1}\subset\mathcal{Q}\subset\mathcal{Q}^*\cap\mathcal{R}=0$, it follows that

$$\dim(\mathcal{R}^*(x))=\sum_{i=1}^{\nu}\dim((\sum_{j\neq i}\mathcal{R}_j+\mathcal{S}_i)(x))-(\nu-1)n\leq\sum_{i=1}^{\nu}\dim(\sum_{j\neq i}\mathcal{R}_j(x))+$$
$$+\sum_{i=1}^{\nu}\dim(\mathcal{S}_i(x))-(\nu-1)n\leq\dim(\mathcal{R}(x))+\dim(\mathcal{Q}^*(x)).$$

The proof of the proposition is complete if we prove that there exists a disjoint partition of $\mathcal{G}$ which satisfies (1.12). For, we will show first that

$$\mathcal{R}_i \cap \mathcal{G} + \mathcal{Q}^* \cap \mathcal{G} = \mathcal{R}_i^* \cap \mathcal{G}, \qquad i = 1, \dots, \nu. \tag{1.16}$$

Indeed, if $\mathcal{Q}^* = 0$, (1.16) trivially follows from proposition 1.3. On the other hand, if $\mathcal{Q}^* \neq 0$, note that, since $\mathcal{R}_i$ and $\mathcal{Q}^*$ are compatible and reasoning as in proposition **3**.3.1, it easily follows that $\mathcal{R}_i + \mathcal{Q}^*$, $i = 1, \dots, \nu$, are controllability distributions. From (1.15) (which can be obtained independently of the fact that (1.12) is true or not) we conclude that $\mathcal{R}_i^* \cap \mathcal{G} + \mathcal{R}^* = \mathcal{R}_i \cap \mathcal{G} + \mathcal{R}^*$. This implies that $\mathcal{R}_i^* \cap \mathcal{G} + \mathcal{Q}^* \cap \mathcal{G} = (\mathcal{R}_i^* \cap \mathcal{G} + \mathcal{R}^*) \cap \mathcal{G} = (\mathcal{R}_i \cap \mathcal{G} + \mathcal{R}^*) \cap \mathcal{G} = \mathcal{R}_i \cap \mathcal{G} + \mathcal{Q}^* \cap \mathcal{G}$, which proves (1.16).

Now, there exist smooth distributions $\mathcal{G}_1^2, \dots, \mathcal{G}_\nu^2, \mathcal{G}_{\nu+1}$ such that

$$\begin{aligned} &\sum_{i=1}^{\nu} (\mathcal{Q}^* \cap \mathcal{R}_i \cap \mathcal{G}) = \mathcal{G}_1^2 \oplus \dots \oplus \mathcal{G}_\nu^2 \oplus \mathcal{G}_{\nu+1}, \\ &\qquad \mathcal{Q}^* \cap \mathcal{R}_i \cap \mathcal{G} = \mathcal{G}_i^2 \oplus \mathcal{G}_{\nu+1}, \qquad i = 1, \dots, \nu, \\ &\qquad\qquad \mathcal{Q} \cap \mathcal{G} = \mathcal{G}_{\nu+1}. \end{aligned} \tag{1.17}$$

Indeed, since $\sum_{i=1}^{\nu} (\mathcal{Q}^* \cap \mathcal{R}_i \cap \mathcal{G}) = \bigcap_{j \neq i} \sum_{k \neq j} (\mathcal{Q}^* \cap \mathcal{R}_k \cap \mathcal{G}) + \sum_{j \neq i} (\mathcal{Q}^* \cap \mathcal{R}_j \cap \mathcal{G})$, it follows from lemma **A**.1 of [40] that there exist smooth distributions $\mathcal{G}_1^2, \dots, \mathcal{G}_\nu^2, \mathcal{G}_{\nu+1}$ such that

$$\begin{aligned} &\sum_{i=1}^{\nu} (\mathcal{Q}^* \cap \mathcal{R}_i \cap \mathcal{G}) = \mathcal{G}_1^2 \oplus \dots \oplus \mathcal{G}_\nu^2 \oplus \mathcal{G}_{\nu+1}, \\ &\sum_{j \neq i} (\mathcal{Q}^* \cap \mathcal{R}_j \cap \mathcal{G}) = \bigoplus_{j \neq i} \mathcal{G}_j^2 \oplus \mathcal{G}_{\nu+1}, \qquad i = 1, \dots, \nu. \end{aligned} \tag{1.18}$$

We claim that $\mathcal{Q}^* \cap \mathcal{R}_i \cap \mathcal{G} = \mathcal{G}_i^2 \oplus \mathcal{G}_{\nu+1}$ and $\mathcal{Q} \cap \mathcal{G} = \mathcal{G}_{\nu+1}$. Indeed, $\mathcal{Q} \cap \mathcal{G} \subset \bigcap_{i=1}^{\nu} \sum_{j \neq i} (\mathcal{Q}^* \cap \mathcal{R}_i \cap \mathcal{G}) = \mathcal{G}_{\nu+1}$. Since $\mathcal{G}_{\nu+1} = \bigcap_{i=1}^{\nu} \sum_{j \neq i} (\mathcal{Q}^* \cap \mathcal{R}_j \cap \mathcal{G}) \subset \bigcap_{i=1}^{\nu} (\sum_{j \neq i} \mathcal{R}_j) \cap (\mathcal{Q}^* \cap \mathcal{G}) = \mathcal{R} \cap \mathcal{Q}^* \cap \mathcal{G} = \mathcal{Q} \cap \mathcal{G}$, we conclude that $\mathcal{G}_{\nu+1} = \mathcal{Q} \cap \mathcal{G}$. On the other hand, $\mathcal{Q}^* \cap \mathcal{R}_i \cap \mathcal{G} \subset \bigcap_{j \neq i} \sum_{k \neq j} (\mathcal{Q}^* \cap \mathcal{R}_k \cap \mathcal{G}) = \mathcal{G}_i^2 \oplus \mathcal{G}_{\nu+1}$ and $\mathcal{G}_{\nu+1} \subset \mathcal{Q}^* \cap \mathcal{R}_i \cap \mathcal{G}$. From (1.18) we obtain $\mathcal{G}_1^2 \oplus \dots \oplus \mathcal{G}_\nu^2 \oplus \mathcal{G}_{\nu+1} = \sum_{i=1}^{\nu} (\mathcal{Q}^* \cap \mathcal{R}_i \cap \mathcal{G}) = (\mathcal{Q}^* \cap \mathcal{R}_i \cap \mathcal{G}) + \sum_{j \notin \{i, \nu+1\}} \mathcal{G}_j^2$. By an easy dimensionality argument, it follows that $\mathcal{G}_i \oplus \mathcal{G}_{\nu+1} = \mathcal{Q}^* \cap \mathcal{R}_i \cap \mathcal{G}$. This proves our claim.

Now, choose a smooth distribution $\mathcal{G}_{\nu+2}$ such that

$$\mathcal{Q}^* \cap \mathcal{G} = \mathcal{G}_1^2 \oplus \dots \oplus \mathcal{G}_\nu^2 \oplus \mathcal{G}_{\nu+1} \oplus \mathcal{G}_{\nu+2},$$

and smooth distributions $\mathcal{G}_1^1, \dots, \mathcal{G}_\nu^1$ such that

$$\mathcal{R}_i \cap \mathcal{G} = \mathcal{G}_i^1 \oplus \mathcal{G}_i^2 \oplus \mathcal{G}_{\nu+1}. \tag{1.19}$$

This, together with (1.16) implies that

$$\mathcal{R}_i^* \cap \mathcal{G} = \mathcal{G}_i^1 + \mathcal{Q}^* \cap \mathcal{G} = \mathcal{G}_i^1 \oplus \mathcal{G}_1^2 \oplus \ldots \oplus \mathcal{G}_\nu^2 \oplus \mathcal{G}_{\nu+1} \oplus \mathcal{G}_{\nu+2}. \tag{1.20}$$

From (1.19) and since $\mathcal{R}_1^*, \ldots, \mathcal{R}_\nu^*$ and $\mathcal{R}_1, \ldots, \mathcal{R}_\nu$ are regular solutions, it follows that

$$\begin{aligned} \mathcal{G} &= \mathcal{G}_1^1 \oplus \ldots \oplus \mathcal{G}_\nu^1 + \mathcal{Q}^* \cap \mathcal{G} = \mathcal{G}_1^1 \oplus \mathcal{G}_1^2 \oplus \ldots \oplus \mathcal{G}_\nu^1 \oplus \mathcal{G}_\nu^2 \oplus \mathcal{G}_{\nu+1} \oplus \mathcal{G}_{\nu+2} = \\ &= (\mathcal{R}_1 \cap \mathcal{G} + \ldots + \mathcal{R}_\nu \cap \mathcal{G}) \oplus \mathcal{G}_{\nu+2}, \end{aligned}$$

which implies $\mathcal{G}_{\nu+2} = 0$, or, equivalently, $\sum_{i=1}^{\nu} (\mathcal{Q}^* \cap \mathcal{R}_i \cap \mathcal{G}) = \mathcal{Q}^* \cap \mathcal{G}$. This, together with (1.17), (1.19) and (1.20), proves (1.12). □

Now, we are ready to prove the following important result. Let (1.9)* and (1.10)* be the dynamics (1.9) and (1.10), respectively associated to the maximal regular solution $\mathcal{R}_1^*, \ldots, \mathcal{R}_\nu^*$.

Proposition 1.9. Assume that $\mathcal{R}_1, \ldots, \mathcal{R}_\nu$ and $\mathcal{R}_1^*, \ldots, \mathcal{R}_\nu^*$ be regular solutions. Moreover, assume that the distribution $\mathcal{Q}^* \cap \mathcal{R}$ has constant dimension in a neighbourhood of x_0. Then, (1.9) is a subdynamics of the corresponding dynamics (1.9), associated to the regular solution $\mathcal{R}_1, \ldots, \mathcal{R}_\nu$. Moreover, the system (1.10), associated to the regular solution $\mathcal{R}_1, \ldots, \mathcal{R}_\nu$, can be obtained from (1.10)* through an invertible dynamic feedback law.* □

Proof. The first part of the proposition can be proved essentially as theorem 3.9 of [40]. Let us plug into (1.8) a regular feedback law $u = \alpha(x) + \beta(x)v$ which renders invariant $\mathcal{R}_1^*, \ldots, \mathcal{R}_\nu^*$ and $\mathcal{R}_1, \ldots, \mathcal{R}_\nu$, with $\beta_j(x)$, $j = 1, \ldots, \nu+1$, chosen in such a way to satisfy (1.13b) and denote by $\tilde{\Sigma}$ the system obtained from (1.8). From standard arguments and proposition 1.8, it follows that there exists a change of coordinates $\eta(x) = (\eta_1^T(x) \;\; \eta_2^T(x) \;\; \eta_3^T(x) \;\; \eta_4^T(x))^T$, $\eta(x_0) = 0$, defined in a neighbourhood of x_0 and such that $\tilde{\Sigma}$ is expressed in these coordinates by

$$\begin{aligned} \dot\eta_1 &= \bar f_{\eta_1}(\eta_1) + \sum_{j=1}^{\nu} \bar g_{\eta_1 j}(\eta_1) u_j, \\ \dot\eta_2 &= \bar f_{\eta_2}(\eta_1, \eta_2) + \sum_{j=1}^{\nu} \bar g_{\eta_2 j}(\eta_1, \eta_2) u_j, \\ \dot\eta_3 &= \bar f_{\eta_3}(\eta_1, \eta_3) + \sum_{j=1}^{\nu} \bar g_{\eta_3 j}(\eta_1, \eta_3) u_j, \\ \dot\eta_4 &= \bar f_{\eta_4}(\eta_1, \ldots, \eta_4) + \sum_{j=1}^{\nu} \bar g_{\eta_4 j}(\eta_1, \ldots, \eta_4) u_j, \\ \dot\eta_5 &= \bar f_{\eta_5}(\eta_1, \ldots, \eta_5) + \sum_{j=1}^{\nu+1} \bar g_{\eta_5 j}(\eta_1, \ldots, \eta_5) u_j, \end{aligned}$$

with

$$
\begin{aligned}
\mathcal{Q} &= \mathrm{span}\{\partial/\partial\eta_5\}, \\
\mathcal{Q}^* \cap \mathcal{R} &= \mathrm{span}\{\partial/\partial\eta_4, \partial/\partial\eta_5\}, \\
\mathcal{Q}^* &= \mathrm{span}\{\partial/\partial\eta_3, \partial/\partial\eta_4, \partial/\partial\eta_5\}, \\
\mathcal{R} &= \mathrm{span}\{\partial/\partial\eta_2, \partial/\partial\eta_4, \partial/\partial\eta_5\}, \\
\mathcal{Q}^* + \mathcal{R} = \mathcal{R}^* &= \mathrm{span}\{\partial/\partial\eta_2\partial/\partial\eta_3, \partial/\partial\eta_4, \partial/\partial\eta_5\}.
\end{aligned}
$$

By definition, the dynamics (1.9) is given by

$$
\begin{aligned}
\dot\eta_2 &= \bar f_{\eta_2}(0, \eta_2), \\
\dot\eta_4 &= \bar f_{\eta_4}(0, \eta_2, 0, \eta_4),
\end{aligned}
\tag{1.21}
$$

and $(1.9)^*$ is given by

$$
\dot\eta_2 = \bar f_{0,\eta_2}(\eta_2). \tag{1.22}
$$

By direct inspection of (1.21) and (1.22), it follows that $(1.9)^*$ is indeed a subdynamics of (1.9).

To prove the second part of the proposition, let us consider the system (1.15), with the matrices $\beta_j(x)$, $j = 1, \dots, \nu+1$, chosen in a such a way to satisfy (1.13b). Note that, in θ–coordinates, we have

$$
\begin{aligned}
\mathcal{R}_i^* + \mathcal{R}^* &= \mathrm{span}\{\partial/\partial\theta_{i1}, \partial/\partial\theta_{j2}, \partial/\partial\theta_{\nu+1} : j = 1, \dots, \nu\}, \\
\mathcal{R}_i + \mathcal{R} &= \mathrm{span}\{\partial/\partial\theta_{ih}, \partial/\partial\theta_{\nu+1} : h = 1, 2\}, \\
\mathcal{R}^* &= \mathrm{span}\{\partial/\partial\theta_{j2}, \partial/\partial\theta_{\nu+1} : j = 1, \dots, \nu\}, \\
\mathcal{R} &= \mathrm{span}\{\partial/\partial\theta_{\nu+1} : j = 1, \dots, \nu\}.
\end{aligned}
$$

By direct inspection of (1.15), it is straightforward to prove that the system (1.10) (after possibly a regular static feedback transformation) is a subsystem of $(1.10)^*$. This, together with proposition 1.7, implies that the system (1.10) can be obtained from $(1.10)^*$ through an invertible dynamic feedback law. □

5.1.2 Necessary and sufficient conditions

From proposition 1.9 it follows that, if NLSS is solvable, the dynamics $(1.9)^*$ must be locally asymptotically stable at the origin. Unfortunately, if NLSS is solvable, one can say at most that the dynamics $(1.10)^*$ is locally asymptotically stabilizable at the origin via *dynamic* feedback. This is a consequence of the fact that, if a nonlinear system (1.1) is *not* asymptotically stabilizable at the origin via *static* feedback, still it may happen to be asymptotically stabilizable at the origin via *dynamic* feedback (see [8]).

However, in some cases the stabilizability properties of (1.10)* can be indeed inferred from those of (1.10). For example, it can be easily shown that, if noninteraction and local *exponential* stability at the origin can be achieved via static feedback laws, (1.10)* must be locally *exponentially* stabilizable at the origin via static feedback (see [52], lemma 7.4.9). Moreover, when $\mathcal{Q}^* = 0$, from proposition 1.3 we obtain $\mathcal{R}_i = \mathcal{R}_i^*$. Thus, if NLSS is solvable, from proposition 1.9 it follows that the dynamics (1.10)* must be locally asymptotically stabilizable at the origin via static feedback.

Nothing can be said, in general, about the dynamics (1.11)*, which may happen not to be asymptotically stabilizable at the origin via static feedback, even if NLSS is solvable. Let us consider the following example

$$\begin{aligned} \dot{x}_i &= u_i, \qquad i = 1, 2, \\ \dot{x}_3 &= x_1 + x_2 - x_3, \\ \dot{x}_4 &= x_4 + x_5^2 - x_2, \\ \dot{x}_5 &= u_3, \\ y_i &= x_i, \qquad i = 1, 2. \end{aligned} \tag{1.23}$$

From straightforward computation, we obtain

$$\mathcal{R}_1^* = \text{span}\{\begin{pmatrix} 1 & 0 & 0 & 0 \\ 0 & 0 & 0 & 0 \\ 0 & 1 & 0 & 0 \\ 0 & 0 & 1 & 0 \\ 0 & 0 & 0 & 1 \end{pmatrix}, \mathcal{R}_2^* = \text{span}\{\begin{pmatrix} 0 & 0 & 0 & 0 \\ 1 & 0 & 0 & 0 \\ 0 & 1 & 0 & 0 \\ 0 & 0 & 1 & 0 \\ 0 & 0 & 0 & 1 \end{pmatrix}\}, \mathcal{Q}^* = \text{span}\{\begin{pmatrix} 0 & 0 \\ 0 & 0 \\ 0 & 0 \\ 1 & 0 \\ 0 & 1 \end{pmatrix}\}.$$

The dynamics (1.11)*, associated to the maximal regular solution, is given by

$$\begin{aligned} \dot{x}_4 &= x_4 + x_5^2, \\ \dot{x}_5 &= u_3, \end{aligned}$$

and it is *not* locally asymptotically stabilizable at the origin via static feedback. Nonetheless, the dynamics (1.11), associated to the regular solution

$$\mathcal{R}_1 = \text{span}\{\begin{pmatrix} 1 & 0 \\ 0 & 0 \\ 0 & 1 \\ 0 & 0 \\ 0 & 0 \end{pmatrix}\}, \mathcal{R}_2 = \mathcal{R}_2^*,$$

is locally asymptotically stabilizable at the origin via static feedback (prove it!). Moreover, the dynamics (1.9) and (1.10), associated to the regular solution $\mathcal{R}_1, \mathcal{R}_2$, are locally exponentially stable at the origin and, respectively, locally exponentially stabilizable at the origin via static feedback. It easily follows that NLSS is solvable. Note that for the linear approximation of (1.23) at $x_0 = 0$ the dynamics (2.2.5) *is* asymptotically stabilizable via static feedback.

On the other hand, if the dynamics (1.9)* is locally asymptotically stable at the origin and, in addition, the dynamics (1.10)* and (1.11)* are locally asymptotically stabilizable at the origin via static feedback laws, it is easy to show that NLSS is solvable.

We sum up all the above results in the following theorem.

Theorem 1.10. Assume that (A1), (A2), (A2b) and (A3) hold and consider the class of regular feedback laws. If

a) the dynamics (1.9) is locally asymptotically stable at the origin,*

b) the dynamics (1.10) is locally asymptotically stabilizable at the origin via static feedback,*

c) the dynamics (1.11) is locally asymptotically stabilizable at the origin via static feedback,*

then NLSS is solvable. Conversely, assume that the distributions $\langle \widetilde{f}, \widetilde{g}_1, \ldots, \widetilde{g}_{\nu+1} | \mathrm{span}\{\widetilde{g}_i, \widetilde{g}_{\nu+1}\}\rangle$, $i = 1, \ldots, \nu$, *have constant dimension in a neighbourhood of* x_0. *If NLSS is solvable, then a) is satisfied. Moreover, if NLSS is solvable with exponential stability, then also b) is satisfied.* □

From proposition 1.3 and theorem 1.10 we obtain the following corollary.

Corollary 1.11. Assume that (A1), (A2), (A2b) and (A3) hold and consider the class of regular feedback laws. Moreover, assume that $Q^* = 0$. *If*

a) the dynamics (1.9) is locally asymptotically stable at the origin,*

b) the dynamics (1.10) is locally asymptotically stabilizable at the origin via static feedback,*

then NLSS is solvable. Conversely, assume that the distributions $\langle \widetilde{f}, \widetilde{g}_1, \ldots, \widetilde{g}_{\nu+1} | \mathrm{span}\{\widetilde{g}_i, \widetilde{g}_{\nu+1}\}\rangle$, $i = 1, \ldots, \nu$, *have constant dimension in a neighbourhood of* x_0. *If NLSS is solvable, then a) and b) are satisfied.* □

Following the proof of theorem 3.3.5, it is easy to prove the following existence condition.

Theorem 1.11. Assume that (A1) and (A3) hold and consider the class of regular feedback laws. If there exists a regular solution $\mathcal{R}_1, \ldots, \mathcal{R}_\nu$ *such that*

a) the dynamics (1.9), associated to the regular solution $\mathcal{R}_1, \ldots, \mathcal{R}_\nu$*, is locally asymptotically stable at the origin,*

b) the dynamics (1.10) and (1.11), associated to the regular solution $\mathcal{R}_1, \ldots, \mathcal{R}_\nu$*, are locally asymptotically stabilizable at the origin via static feedback,*

then NLSS is solvable. Conversely, if NLSS is solvable and if, in addition, the set of distributions $\mathcal{R}_i = \langle \widetilde{f}, \widetilde{g}_1, \ldots, \widetilde{g}_{\nu+1} | \mathrm{span}\{\widetilde{g}_i, \widetilde{g}_{\nu+1}\}\rangle$, $i = 1, \ldots, \nu$, *is regular, then* $\mathcal{R}_1, \ldots, \mathcal{R}_\nu$ *is a regular solution, which satisfies a) and b).* □

Clearly, conditions a) and b) of theorem 1.11 cannot be checked directly, since we do not know *a priori* the family of regular solutions.

5.2 Noninteracting control with local stability via dynamic feedback

In this section we will give some necessary and sufficient conditions to solve NLSD for the class of systems, which satisfy assumptions (A1), (A2), (A2b) and (A3). Moreover, we will consider the class of invertible feedback laws. Although there is no practical reason for so costraining the class of systems and feedback laws considered, this allows us to extend significantly the results of section 3.4.

Before going further in our analysis, some generalization are in order. This is done in the next section.

5.2.1 Invariant dynamics induced by a regular solution

In analogy with section 3.1, to a given system (1.1), which satisfies assumptions (A1), (A2), (A2b) and (A3), and to a given regular solution $\mathcal{R}_1, \ldots, \mathcal{R}_\nu$, we can *uniquely* associate (in some sense to be specified) certain nonlinear dynamics, which will be crucial in solving NLSD for the class of systems considered.

As mentioned above, we will begin our analysis under assumptions (A1), (A2), (A2b) and (A3). Given a regular solution $\mathcal{R}_1, \ldots, \mathcal{R}_\nu$, we can suppose that (1.1) is already in the form (1.8) (for simplicity, we let $z = x$ and $v = u$). Also, for simplicity of notation, we will assume that dim $(\mathrm{span}\{\widetilde{g}_i\}(x)) = 1$ for $i = 1, \ldots, \nu + 1$ (the general case goes in the same way).

Now, let us introduce the following distribution. Let $\mathcal{I}$ be the Lie ideal generated by the vector fields $\{\widetilde{g}_{\nu+1}, [\widetilde{g}_i, ad^k_{\widetilde{f}}\widetilde{g}_j] \,:\, k \geq 0,\ i,j = 1,\ldots,\nu \text{ and } i \neq j\}$ in the Lie algebra generated by $\widetilde{f}, \widetilde{g}_1, \ldots, \widetilde{g}_{\nu+1}$ and let

$$\Delta_{MIX} = \operatorname{span}\{\tau : \tau \in \mathcal{I}\}. \tag{2.1}$$

It is not hard to show that $\Delta_{MIX} \subset \mathcal{R}$ (see section 3.4.1) and $\mathcal{Q} \subset \Delta_{MIX}$ (see Figure 2.1). In few words, Δ_{MIX} is spanned by the Lie brackets of $\widetilde{f}, \widetilde{g}_1, \ldots, \widetilde{g}_{\nu+1}$, in which appear either $\widetilde{g}_{\nu+1}$ or both $\widetilde{g}_i$ and $\widetilde{g}_j$ with $i,j \in \{1,\ldots,\nu\}$ and $i \neq j$. When $m = p = \nu$, we obtain the distribution Δ_{MIX}, as defined in section **3**.4.1. Moreover, in the case of linear systems, $\Delta_{MIX} = \mathcal{Q}$. Let us assume the following.

(A4b) The distributions Δ_{MIX} and $\Delta_{MIX} + \mathcal{R}_i$, $i = 1,\ldots,\nu$ have constant dimension for all x in a neighbourhood of x_0.

In [3] the distribution Δ_{MIX}, associated to the maximal regular solution $\mathcal{R}_1^*, \ldots, \mathcal{R}_\nu^*$, has been defined in a different way. It can be shown that, under our assumptions, the distribution (2.1), associated to $\mathcal{R}_1^*, \ldots, \mathcal{R}_\nu^*$, and the distribution $\Delta_{MIX}^* + \mathcal{Q}^*$, as defined in [3], do coincide. Here, for simplicity of notation, we prefer the definition (2.1).

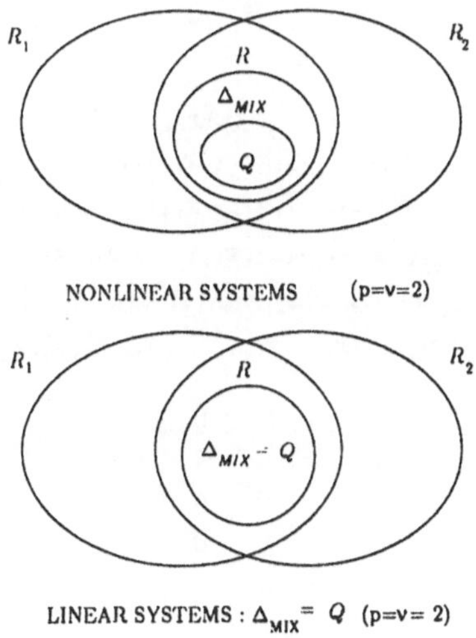

Figure 2.1

At this point, the question may arise if Δ_{MIX}, as defined in (2.1), is indeed the natural substitute of the corresponding distribution defined in section **3**.4.1. The fact that in the linear case $\Delta_{MIX} = \mathcal{Q}$ suggests that we have to look for something smaller than Δ_{MIX}. Toward this end, we have to dig further.

In what follows, we assume that any regular set $\mathcal{R}_1, \dots, \mathcal{R}_\nu$ satisfies also assumption (A4b). In analogy with section 3.4.1, we can easily prove the following result.

Proposition 2.1. Assume that (A1) and (A3) hold and let $\mathcal{R}_1, \dots, \mathcal{R}_\nu$ be a regular solution. Then, the distributions Δ_{MIX}, $\Delta_{MIX}+\mathcal{R}_i$ and $(\Delta_{MIX}+\mathcal{R}_i)\cap\mathcal{R}$, $i = 1, \dots, m$, have constant dimension in a neighbourhood of x_0, are involutive and invariant under $\widetilde{f}, \widetilde{g}_1, \dots, \widetilde{g}_{\nu+1}$. □

Following section 3.4.1, it is not hard to show that there exists a change of coordinates

$$z^i(x) = \begin{pmatrix} x_1^T & \cdots & x_{i-1}^T & x_{i+1}^T & \cdots & x_m^T & (z_1^i(x))^T & \cdots & (z_5^i(x))^T \end{pmatrix}^T, \tag{2.2}$$

$z^i(x_0) = 0$, defined in a neighborhood of x_0 and such that the vector fields $\widetilde{f}$ and $\widetilde{g}_i$ are expressed in these coordinates by

$$\widetilde{f}(z^i) = \begin{pmatrix} \widetilde{f}_1(x_1) \\ \vdots \\ \widetilde{f}_{i-1}(x_{i-1}) \\ \widetilde{f}_{i+1}(x_{i-1}) \\ \vdots \\ \widetilde{f}_m(x_m) \\ \widetilde{f}_{z_1^i}(x_1, \dots, x_{i-1}, x_{i+1}, \dots, x_m, z_1^i) \\ \widetilde{f}_{z_2^i}(x_1, \dots, x_{i-1}, x_{i+1}, \dots, x_m, z_2^i) \\ \widetilde{f}_{z_3^i}(x_1, \dots, x_{i-1}, x_{i+1}, \dots, x_m, z_1^i, z_2^i, z_3^i) \\ \widetilde{f}_{z_4^i}(x_1, \dots, x_{i-1}, x_{i+1}, \dots, x_m, z_1^i, z_2^i, z_3^i, z_4^i) \\ \widetilde{f}_{z_5^i}(x_1, \dots, x_{i-1}, x_{i+1}, \dots, x_m, z_1^i, z_2^i, z_3^i, z_4^i, z_5^i) \end{pmatrix},$$

$$\widetilde{g}_i(z^i) = \begin{pmatrix} 0 \\ \vdots \\ 0 \\ 0 \\ \vdots \\ 0 \\ \widetilde{g}_{iz_1^i}(x_1, \dots, x_{i-1}, x_{i+1}, \dots, x_m, z_1^i) \\ 0 \\ \widetilde{g}_{iz_3^i}(x_1, \dots, x_{i-1}, x_{i+1}, \dots, x_m, z_1^i, z_2^i, z_3^i) \\ \widetilde{g}_{iz_4^i}(x_1, \dots, x_{i-1}, x_{i+1}, \dots, x_m, z_1^i, z_2^i, z_3^i, z_4^i) \\ \widetilde{g}_{iz_5^i}(x_1, \dots, x_{i-1}, x_{i+1}, \dots, x_m, z_1^i, z_2^i, z_3^i, z_4^i, z_5^i) \end{pmatrix},$$

with

$$
\begin{aligned}
\mathcal{R}_i + \mathcal{R} &= span\{\partial/\partial z_1^i, \partial/\partial z_2^i, \partial/\partial z_3^i, \partial/\partial z_4^i, \partial/\partial z_5^i\}, \\
\mathcal{R}_i + \Delta_{MIX} &= span\{\partial/\partial z_1^i, \partial/\partial z_3^i, \partial/\partial z_4^i, \partial/\partial z_5^i\}, \\
\mathcal{R} &= span\{\partial/\partial z_2^i, \partial/\partial z_3^i, \partial/\partial z_4^i, \partial/\partial z_5^i\}, \\
\mathcal{R} \cap (\mathcal{R}_i + \Delta_{MIX}) &= span\{\partial/\partial z_3^i, \partial/\partial z_4^i, \partial/\partial z_5^i\}, \\
\Delta_{MIX} &= span\{\partial/\partial z_4^i, \partial/\partial z_5^i\}, \\
\mathcal{Q} &= span\{\partial/\partial z_5^i\}.
\end{aligned}
$$

Consider now the noninteractive system (1.8). In analogy with section **3**.4.1, it makes sense to consider the restriction $\widetilde{f}|\mathcal{L}_{x_0}^{\Delta_{MIX}}$. This vector field induces a dynamics evolving on $\mathcal{L}_{x_0}^{\Delta_{MIX}}$, which in z^i–coordinates is described by

$$
\begin{aligned}
\dot{z}_4^i &= \widetilde{f}_{z_4^i}(0, \ldots, 0, z_4^i), \\
\dot{z}_5^i &= \widetilde{f}_{z_5^i}(0, \ldots, 0, z_4^i, z_5^i).
\end{aligned}
$$

Let us consider the subdynamics

$$
\dot{z}_4^i = \widetilde{f}_{z_4^i}(0, \ldots, 0, z_4^i). \tag{2.3}
$$

The dynamics (2.3) is trivial in the case of linear systems, since $\Delta_{MIX} = \mathcal{Q}$.

It makes sense to consider also the restrictions $\widetilde{f}|\mathcal{L}_{x_0}^{\mathcal{R}_i+\Delta_{MIX}}$ and $\widetilde{g}_i|\mathcal{L}_{x_0}^{\mathcal{R}_i+\Delta_{MIX}}$ (the latter since $\widetilde{g}_i \in \mathcal{R}_i$). These vector fields induces a control system evolving on $\mathcal{L}_{x_0}^{\mathcal{R}_i+\Delta_{MIX}}$, described in z^i–coordinates by

$$
\begin{pmatrix} \dot{z}_1^i \\ \dot{z}_3^i \\ \dot{z}_4^i \\ \dot{z}_5^i \end{pmatrix} = \begin{pmatrix} \widetilde{f}_{z_1^i}(0, \ldots, 0, z_1^i) \\ \widetilde{f}_{z_3^i}(0, \ldots, 0, z_1^i, 0, z_3^i) \\ \widetilde{f}_{z_4^i}(0, \ldots, 0, z_1^i, 0, z_3^i, z_4^i) \\ \widetilde{f}_{z_5^i}(0, \ldots, 0, z_1^i, 0, z_3^i, z_4^i, z_5^i) \end{pmatrix} + \begin{pmatrix} \widetilde{g}_{iz_1^i}(0, \ldots, 0, z_1^i) \\ \widetilde{g}_{iz_3^i}(0, \ldots, 0, z_1^i, 0, z_3^i) \\ \widetilde{g}_{iz_4^i}(0, \ldots, 0, z_1^i, 0, z_3^i, z_4^i) \\ \widetilde{g}_{iz_5^i}(0, \ldots, 0, z_1^i, 0, z_3^i, z_4^i, z_5^i) \end{pmatrix} u_i.
$$

Let us consider the subsystem

$$
\begin{pmatrix} \dot{z}_1^i \\ \dot{z}_3^i \end{pmatrix} = \begin{pmatrix} \widetilde{f}_{z_1^i}(0, \ldots, 0, z_1^i) \\ \widetilde{f}_{z_3^i}(0, \ldots, 0, z_1^i, 0, z_3^i) \end{pmatrix} + \begin{pmatrix} \widetilde{g}_{iz_1^i}(0, \ldots, 0, z_1^i) \\ \widetilde{g}_{iz_3^i}(0, \ldots, 0, z_1^i, 0, z_3^i) \end{pmatrix} u_i. \tag{2.4}
$$

It is worth noting that, in the case of linear systems, (2.4) is a controllable system, since $\Delta_{MIX} = \mathcal{Q}$ and $\mathcal{R}_i$ is a controllability subspace.

A first question arises if the distribution Δ_{MIX}, associated to the regular solution $\mathcal{R}_1, \ldots, \mathcal{R}_\nu$, is invariant under regular feedback laws, which render invariant $\mathcal{R}_1, \ldots, \mathcal{R}_\nu$. This is answered in the next proposition, which can be proven exactly as proposition

3.4.2. Given a feedback law $u = \alpha'(x) + \beta'(x)v$, let us partition $\alpha'(x)$ and $\beta'(x)$ as $\alpha'(x) = ((\alpha_1')^T(x) \quad \cdots \quad (\alpha_{\nu+1}')^T(x))^T$ and $\beta'(x) = (\beta_1'(x) \quad \cdots \quad \beta_{\nu+1}'(x))$, respectively. Moreover, let $\widetilde{\Sigma}$ be the system (1.8) corresponding to the given regular solution $\mathcal{R}_1, \ldots, \mathcal{R}_\nu$ and denote by $\widetilde{\Sigma}'$ the closed–loop system resulting from plugging this feedback law into $\widetilde{\Sigma}$. Correspondingly, let us denote by Δ'_{MIX} the distribution defined as Δ_{MIX} but with $\widetilde{f}$ and $\widetilde{g}_j$, $j = 1, \ldots, \nu + 1$, replaced by $\widetilde{f}' = \widetilde{f} + \widetilde{G}\alpha'$ and $\widetilde{g}_j' = \widetilde{G}\beta_j'$, respectively.

Proposition 2.2. Assume that (A1) and (A3) hold and let $\mathcal{R}_1, \ldots, \mathcal{R}_\nu$ be a regular solution. Let $u = \alpha'(x) + \beta'(x)v$ be a regular feedback law, which satisfies (1.7). We have $\Delta_{MIX} = \Delta'_{MIX}$ for all x in a neighbourhood of x_0. Moreover, each component of $\alpha_i'(x)$ is equal to 0 for $x \in \mathcal{L}_{x_0}^{\mathcal{R}_j + \Delta_{MIX}}$, $j \neq i$ and $i = 1, \ldots, \nu$; in particular, each component of $\alpha'(x)$ is equal to 0 for $x \in \mathcal{L}_{x_0}^{\Delta_{MIX}}$. □

By proposition 2.2 and reasoning as in proposition 1.7, we can also prove the following. Let $\widetilde{\Sigma}$ and $\widetilde{\Sigma}'$ be as above. Moreover, let (2.3)′ and (2.4)′ be the dynamics defined on $\widetilde{\Sigma}'$ and corresponding to (2.3) and (2.4), respectively.

Proposition 2.3. The dynamics (2.3) and (2.3)′ are equal up to changes of coordinates, defined in a neighbourhood of x_0. The systems (2.4) and (2.4)′ are equal up to changes of coordinates and invertible static feedback transformations. □

While proposition 2.2 solves our doubts about the invariance of the distribution Δ_{MIX}, associated to a regular solution $\mathcal{R}_1, \ldots, \mathcal{R}_\nu$, under regular feedback laws, which render invariant $\mathcal{R}_1, \ldots, \mathcal{R}_\nu$, we are left wondering if there is any significant relation between the dynamics (2.3) (or (2.4)), respectively associated to *different* regular solutions. We will show that the dynamics (2.3), associated to the maximal regular solution $\mathcal{R}_1^*, \ldots, \mathcal{R}_\nu^*$, is a *subdynamics* of the corresponding dynamics (2.3), associated to any other regular solution $\mathcal{R}_1, \ldots, \mathcal{R}_\nu$. On the other hand, we will show also that the system (2.4), associated to a given regular solution $\mathcal{R}_1, \ldots, \mathcal{R}_\nu$, can be obtained from the corresponding system (2.4), associated to the maximal regular solution $\mathcal{R}_1^*, \ldots, \mathcal{R}_\nu^*$, through an *invertible* dynamic feedback law.

Toward this end, we need some preliminary results. Let us denote by Δ_{MIX} and Δ^*_{MIX}, respectively, the distribution (2.1) associated to the regular solution $\mathcal{R}_1, \ldots, \mathcal{R}_\nu$ and the corresponding distribution (2.1) associated to the maximal regular solution $\mathcal{R}_1^*, \ldots, \mathcal{R}_\nu^*$.

Proposition 2.4. Suppose that $\mathcal{R}_1, \ldots, \mathcal{R}_\nu$ and $\mathcal{R}_1^*, \ldots, \mathcal{R}_\nu^*$ be regular solution. Then, if the distribution $\Delta_{MIX} \cap \mathcal{Q}^*$ has constant dimension in a neighbourhood of x_0, we have

$$\Delta_{MIX} + \mathcal{Q}^* = \Delta^*_{MIX}.$$

Proof. From the proof of proposition 1.8 we know that there exist a regular feedback law $u = \alpha(x) + \beta(x)v$ and a block partition $\beta_1^1(x), \beta_1^2(x), \ldots, \beta_\nu^1(x), \beta_\nu^2(x), \beta_{\nu+1}(x)$ of $\beta(x)$ which satisfy (1.12) and (1.13). Let $\widetilde{\Sigma}$ be the system obtained from plugging this feedback law into (1.8). Moreover, let $f + G\alpha = \widetilde{f}$, $G\beta_i^j = \widetilde{g}_i^j$, $\mathcal{G}_i^j = \text{span}\{\widetilde{g}_i^j\}$ and $\beta_i(x) = (\beta_i^1(x) \quad \beta_i^2(x))$ for $j = 1,2$ and $i = 1, \ldots, \nu$, $G\beta_i = \widetilde{g}_i$ for $i = 1, \ldots, \nu + 1$, $\beta_{\nu+1}^1(x) = (\beta_1^2(x) \quad \cdots \quad \beta_\nu^2(x) \quad \beta_{\nu+1}(x))$ and $G\beta_{\nu+1}^1 = \widetilde{g}_{\nu+1}^1$.

Since Δ_{MIX} and Δ_{MIX}^* are invariant under regular feedback laws, which render invariant the distributions $\mathcal{R}_1^*, \ldots, \mathcal{R}_\nu^*$ and $\mathcal{R}_1, \ldots, \mathcal{R}_\nu$, respectively, we can compute these distributions directly on $\widetilde{\Sigma}$. By definition, Δ_{MIX} is spanned by the vector fields of the Lie ideal generated by $\{\widetilde{g}_{\nu+1}, [\widetilde{g}_i, ad_{\widetilde{f}}^k \widetilde{g}_j] : k \geq 0,\ i,j = 1, \ldots, \nu \text{ and } i \neq j\}$ in the Lie algebra generated by $\widetilde{f}, \widetilde{g}_1, \ldots, \widetilde{g}_{\nu+1}$. On the other hand, Δ_{MIX}^* is spanned by the vector fields of the Lie ideal generated by $\{\widetilde{g}_{\nu+1}^1, [\widetilde{g}_i^1, ad_{\widetilde{f}}^k \widetilde{g}_j^1] : k \geq 0,\ i,j = 1, \ldots, \nu \text{ and } i \neq j\}$ in the Lie algebra generated by $\widetilde{f}, \widetilde{g}_1^1, \ldots, \widetilde{g}_\nu^1, \widetilde{g}_{\nu+1}^1$. Clearly, by definition it follows that $\Delta_{MIX} + \mathcal{Q}^* \subset \Delta_{MIX}^*$. We claim that also $\Delta_{MIX}^* \subset \Delta_{MIX} + \mathcal{Q}^*$. Indeed, $\mathcal{G}_1^2 \oplus \ldots \oplus \mathcal{G}_\nu^2 \oplus \mathcal{G}_{\nu+1} \subset \mathcal{Q}^*$. Since $\mathcal{Q}^*$ is invariant under $\widetilde{f}, \widetilde{g}_1, \ldots, \widetilde{g}_{\nu+1}$, it is not hard to see that the vector fields, which are in Δ_{MIX}^* but *not* in Δ_{MIX}, are contained in $\mathcal{Q}^*$. This proves our claim and the proposition.□

Now, we are ready to state the main result of this section. Let (2.3)* and (2.4)* be the dynamics (2.3) and (2.4), respectively associated to the maximal regular solution $\mathcal{R}_1^*, \ldots, \mathcal{R}_\nu^*$.

Proposition 2.5. Let $\mathcal{R}_1, \ldots, \mathcal{R}_\nu$ and $\mathcal{R}_1^*, \ldots, \mathcal{R}_\nu^*$ be regular solutions. Moreover, assume that the distribution $\Delta_{MIX} \cap \mathcal{Q}^*$ has constant dimension in a neighbourhood of x_0. Then, the dynamics (2.3) is a subdynamics of the dynamics (2.3). Moreover, the system (2.4) can be obtained from (2.4)* through an invertible dynamic feedbak law.* □

Proof. Let $\widetilde{\Sigma}$ be as above. The distributions $\mathcal{Q}^*$, $\mathcal{Q}$, $\Delta_{MIX} + \mathcal{Q}^*$, $\Delta_{MIX} \cap \mathcal{Q}^*$, Δ_{MIX} and Δ_{MIX}^* have constant dimension in a neighbourhood of x_0, are involutive and invariant under $\widetilde{f}, \widetilde{g}_1, \ldots, \widetilde{g}_{\nu+1}$. By Frobenius' theorem and proposition 2.4, there exists a change of coordinates $\zeta(x) = (x_1^T \ \cdots x_\nu^T \quad \zeta_{\nu+1}^T(x) \quad \cdots \quad \zeta_{\nu+5}^T(x))^T$, $\zeta(x_0) = 0$, defined in a neighbourhood of x_0, such that $\widetilde{\Sigma}$ is expressed in these coordinates by

$$\begin{aligned}
\dot{x}_i &= \widetilde{f}_i(x_i) + \widetilde{g}_{ii}(x_i)u_i, \qquad i = 1, \ldots, \nu, \\
\dot{\zeta}_{\nu+1} &= \widetilde{f}_{\zeta_{\nu+1}}(x_1, \ldots, x_\nu, \zeta_{\nu+1}) + \sum_{j=1}^{\nu} \widetilde{g}_{\zeta_{\nu+1},j}(x_1, \ldots, x_\nu, \zeta_{\nu+1})u_j, \\
\dot{\zeta}_{\nu+2} &= \widetilde{f}_{\zeta_{\nu+2}}(x_1, \ldots, x_\nu, \zeta_{\nu+1}, \zeta_{\nu+2}) + \sum_{j=1}^{\nu} \widetilde{g}_{\zeta_{\nu+2},j}(x_1, \ldots, x_\nu, \zeta_{\nu+1}, \zeta_{\nu+2})u_j, \\
\dot{\zeta}_{\nu+3} &= \widetilde{f}_{\zeta_{\nu+3}}(x_1, \ldots, x_\nu, \zeta_{\nu+1}, \zeta_{\nu+3}) + \sum_{j=1}^{\nu} \widetilde{g}_{\zeta_{\nu+1},j}(x_1, \ldots, x_\nu, \zeta_{\nu+1}, \zeta_{\nu+3})u_j, \\
\dot{\zeta}_{\nu+4} &= \widetilde{f}_{\zeta_{\nu+4}}(x_1, \ldots, x_\nu, \zeta_{\nu+1}, \ldots, \zeta_{\nu+4}) + \sum_{j=1}^{\nu} \widetilde{g}_{\zeta_{\nu+4},j}(x_1, \ldots, x_\nu, \zeta_{\nu+1}, \ldots, \zeta_{\nu+4})u_j, \\
\dot{\zeta}_{\nu+5} &= \widetilde{f}_{\zeta_{\nu+5}}(x_1, \ldots, x_\nu, \zeta_{\nu+1}, \ldots, \zeta_{\nu+5}) + \sum_{j=1}^{\nu+1} \widetilde{g}_{\zeta_{\nu+5},j}(x_1, \ldots, x_\nu, \zeta_{\nu+1}, \ldots, \zeta_{\nu+5})u_j,
\end{aligned}$$

with

$$\begin{aligned}
\mathcal{Q} &= \mathrm{span}\{\partial/\partial\zeta_{\nu+5}\}, \\
\Delta_{MIX} \cap \mathcal{Q}^* &= \mathrm{span}\{\partial/\partial\zeta_{\nu+4}, \partial/\partial\zeta_{\nu+5}\}, \\
\Delta_{MIX} &= \mathrm{span}\{\partial/\partial\zeta_{\nu+3}, \partial/\partial\zeta_{\nu+4}, \partial/\partial\zeta_{\nu+5}\}, \\
\mathcal{Q}^* &= \mathrm{span}\{\partial/\partial\zeta_{\nu+2}, \partial/\partial\zeta_{\nu+4}, \partial/\partial\zeta_{\nu+5}\}, \\
\Delta^*_{MIX} &= \mathrm{span}\{\partial/\partial\zeta_{\nu+2}, \partial/\partial\zeta_{\nu+3}, \partial/\partial\zeta_{\nu+4}, \partial/\partial\zeta_{\nu+5}\}, \\
\mathcal{R}^* &= \mathrm{span}\{\partial/\partial\zeta_{\nu+1}, \partial/\partial\zeta_{\nu+2}, \partial/\partial\zeta_{\nu+3}, \partial/\partial\zeta_{\nu+4}, \partial/\partial\zeta_{\nu+5}\},
\end{aligned}$$

and the partition $u_1, \ldots, u_{\nu+1}$ of the input vector u corresponds to the partition $\beta_1(x), \ldots,$ $\beta_{\nu+1}(x)$ of $\beta(x)$ in proposition 2.4.

Clearly, in ζ–coordinates, the dynamics (2.3), associated to the regular solution $\mathcal{R}_1, \ldots,$ $\mathcal{R}_\nu$, is given by

$$\begin{aligned}
\dot{\zeta}_{\nu+3} &= \widetilde{f}_{\zeta_{\nu+3}}(0, \ldots, 0, \zeta_{\nu+3}), \\
\dot{\zeta}_{\nu+4} &= \widetilde{f}_{\zeta_{\nu+4}}(0, \ldots, 0, \zeta_{\nu+3}, \zeta_{\nu+4}),
\end{aligned} \tag{2.5}$$

and the dynamics (2.3), associated to the maximal regular solution $\mathcal{R}_1^*, \ldots, \mathcal{R}_\nu^*$, is given by

$$\dot{\zeta}_{\nu+3} = \widetilde{f}_{\zeta_{\nu+3}}(0, \ldots, 0, \zeta_{\nu+3}). \tag{2.6}$$

By direct inspection of (2.5) and (2.6), it follows that (2.3)* is a subdynamics of (2.3).

Now, we will prove the second part of the proposition. Let $u_1^1, u_1^2, \ldots, u_\nu^1, u_\nu^2, u_{\nu+1}$ be the partition of the input vector u corresponding to the partition $\beta_1^1(x), \beta_1^2(x), \ldots, \beta_\nu^1(x), \beta_\nu^2$ $(x), \beta_{\nu+1}(x)$ of $\beta(x)$ (see the proof of proposition 1.8).

The distributions Δ_{MIX}, Δ^*_{MIX}, $\Delta_{MIX} \cap \mathcal{Q}^*$, $\Delta^*_{MIX} \cap (\mathcal{R}_i + \Delta_{MIX})$, $\mathcal{R}_i + \Delta_{MIX}$ and $\mathcal{R}^*_i + \Delta^*_{MIX}$ have constant dimension in a neighbourhood of x_0, are involutive and invariant under $\tilde{f}, \tilde{g}_1, \ldots, \tilde{g}_{\nu+1}$. Moreover, $\mathcal{G}^2_1 \oplus \ldots \oplus \mathcal{G}^2_\nu \oplus \mathcal{G}_{\nu+1} \subset \mathcal{Q}^*$. By Frobenius' theorem and proposition 2.4, there exists a change of coordinates $\xi(x) = (\,\xi_1^T(x) \quad \cdots \quad \xi_5^T(x)\,)^T$, $\xi(x_0) = 0$, defined in a neighbourhood of x_0 and such that $\tilde{\Sigma}$ is expressed in these coordinates by

$$
\begin{aligned}
\dot{\xi}_1 &= \tilde{f}_{\xi_1}(\xi_1) + \sum_{j \neq i} \tilde{g}^1_{\xi_1,j}(\xi_1) u^1_j, \\
\dot{\xi}_2 &= \tilde{f}_{\xi_2}(\xi_1, \xi_2) + \sum_{j=1}^{\nu} \tilde{g}^1_{\xi_2,j}(\xi_1, \xi_2) u^1_j, \\
\dot{\xi}_3 &= \tilde{f}_{\xi_3}(\xi_1, \xi_3) + \sum_{h=1}^{2} \sum_{j=1}^{\nu} \tilde{g}^h_{\xi_3,j}(\xi_1, \xi_3) u^h_j + \tilde{g}_{\xi_3,\nu+1}(\xi_1, \xi_3) u_{\nu+1}, \\
\dot{\xi}_4 &= \tilde{f}_{\xi_4}(\xi_1, \ldots, \xi_4) + \sum_{h=1}^{2} \sum_{j=1}^{\nu} \tilde{g}^h_{\xi_4,j}(\xi_1, \ldots, \xi_4) u^h_j + \tilde{g}_{\xi_4,\nu+1}(\xi_1, \ldots, \xi_4) u_{\nu+1}, \\
\dot{\xi}_5 &= \tilde{f}_{\xi_5}(\xi_1, \ldots, \xi_5) + \sum_{h=1}^{2} \sum_{j=1}^{\nu} \tilde{g}^h_{\xi_5,j}(\xi_1, \ldots, \xi_5) u^h_j + \tilde{g}_{\xi_5,\nu+1}(\xi_1, \ldots, \xi_5) u_{\nu+1},
\end{aligned}
$$

where

$$
\begin{aligned}
\Delta_{MIX} &= \text{span}\{\partial/\partial\xi_5\}, \\
\Delta^*_{MIX} \cap (\mathcal{R}_i + \Delta_{MIX}) &= \text{span}\{\partial/\partial\xi_4, \partial/\partial\xi_5\}, \\
\Delta^*_{MIX} &= \text{span}\{\partial/\partial\xi_3, \partial/\partial\xi_4, \partial/\partial\xi_5, \}, \\
\mathcal{R}_i + \Delta_{MIX} &= \text{span}\{\partial/\partial\xi_2, \partial/\partial\xi_4, \partial/\partial\xi_5\}, \\
\mathcal{R}^*_i + \Delta^*_{MIX} &= \text{span}\{\partial/\partial\xi_2, \partial/\partial\xi_3, \partial/\partial\xi_4, \partial/\partial\xi_5\}.
\end{aligned}
$$

Clearly, in ξ-coordinates, the system (2.4), associated to the regular solution $\mathcal{R}_1, \ldots, \mathcal{R}_\nu$, is given by

$$
\begin{aligned}
\dot{\xi}_2 &= \tilde{f}_{\xi_2}(0, \xi_2) + \tilde{g}^1_{\xi_2,i}(0, \xi_2) u^1_i, \\
\dot{\xi}_4 &= \tilde{f}_{\xi_4}(0, \xi_2, 0, \xi_4) + \sum_{h=1}^{2} \tilde{g}^h_{\xi_4,j}(0, \xi_2, 0, \xi_4) u^h_i.
\end{aligned} \tag{2.7}
$$

On the other hand, the system $(2.6)^*$, associated to the maximal regular solution $\mathcal{R}^*_1, \ldots, \mathcal{R}^*_\nu$, is given by

$$
\dot{\xi}_2 = \tilde{f}_{\xi_2}(0, \xi_2) + \tilde{g}^1_{\xi_2,i}(0, \xi_2) u^1_i. \tag{2.8}
$$

By direct inspection of (2.7) and (2.8), it follows that the system $(2.4)^*$, associated to the maximal regular solution $\mathcal{R}^*_1, \ldots, \mathcal{R}^*_\nu$, is a subdynamics of the system (2.4), associated to the regular solution $\mathcal{R}_1, \ldots, \mathcal{R}_\nu$. This, together with proposition 2.3, implies that, the dynamics (2.4) can be obtained from $(2.4)^*$ through an invertible dynamic feedback law.□

5.2.2 Necessary and sufficient conditions for a class of nonlinear systems

In this section, we will give some necessary and sufficient conditions to solve NLSD for the class of systems considered in section 2.1. Let us begin with necessary conditions. Assume that NLSD is solvable and consider the class of invertible feedback laws. What can be said about the dynamics (1.9)* and (1.10)*, respectively? We will show that, in analogy with the case $m = p = \nu$, if NLSD is solvable, then the dynamics (1.9)* must be locally asymptotically stable at the origin and the dynamics (1.10)* must be locally asymptotically stabilizable at the origin via dynamic feedback. On the other hand, as we have shown in the example at the beginning of section 1.1, if NLSD is solvable, the dynamics (1.11)* is *not* necessarily asymptotically stabilizable at the origin.

Before going further, we need the following definition, which generalizes the one given in section **3**.1.

Definition 2.6. An invertible feedback law

$$\begin{aligned} u &= \alpha(x,w) + \beta(x,w)v, \\ \dot{w} &= \delta(x,w) + \gamma(x,w)v, \qquad w \in I\!R^{n^w}, \end{aligned} \tag{2.9}$$

is *weakly noninteractive* (in a generalized sense) if there exist $\mu \geq 1$ and block partitions

$$\alpha(x,w) = (\, \alpha_1^T(x,w) \quad \cdots \quad \alpha_{\mu+1}^T(x,w)\,)^T\,,\ \beta(x,w) = (\, \beta_1(x,w) \quad \cdots \quad \beta_{\mu+1}(x,w)\,)\,,$$

$$\beta_j(x,w) = (\, \beta_{1j}^T(x,w) \quad \cdots \quad \beta_{\mu+1,j}^T(x,w)\,)^T\,, \qquad j = 1,\dots,\mu+1,$$

such that for all x^e in a neighbourhood of x_0^e

$$\beta_{ij}(x^e) = 0, \qquad j \neq i,\, i \neq \mu+1, \tag{2.10}$$

$$L_{\widetilde{g}_j^e}\widetilde{D}^e\alpha_i(x^e) = 0, \qquad j \neq i,\, i \neq \mu+1, \tag{2.11}$$

$$L_{\widetilde{g}_j^e}\widetilde{D}^e\beta_{ii}(x^e) = 0, \qquad j \neq i,\, i \neq \mu+1, \tag{2.12}$$

where $\widetilde{D}^e$ is an arbitrary composition of the Lie derivatives $L_{\widetilde{f}^e}$ and $L_{\widetilde{g}_j^e}$, $j = 1,\dots,\mu+1$.□

When $m = p = \nu$, the above definition coincides with the one given in section **3**.1 (in this case, we have $\mu = m$).

The following lemma generalizes corollary **3**.1.4. Moreover, it proves that every feedback law, which solves ND for the class of systems considered, can be expressed as the cascade of a *regular* feedback law together with a *weakly invertible* feedback law. Let $\widetilde{\Sigma}^*$ be the system (1.8) with $\mathcal{R}_i = \mathcal{R}_i^*$.

Lemma 2.7. Assume that (A1), (A2), (A2b) and (A3) hold. Any invertible feedback law, which solves ND for $\widetilde{\Sigma}^*$, is weakly noninteractive (in a generalized sense). □

Proof. Let $\widetilde{\Sigma}^e$ be the system resulting from plugging an invertible feedback law, which solves ND, into $\widetilde{\Sigma}^*$. For simplicity of notation, we will assume that dim (span $\{\widetilde{g}_i^e\})(x^e)) = 1$ for $i = 1, \ldots, \nu+1$.

Since $\widetilde{\Sigma}^e$ is noninteractive, from lemma 3.3.1 of [52] it follows that

$$L_{\widetilde{g}_j^e}\widetilde{D}^e h_i^e(x^e) = 0, \qquad i \neq j, \tag{2.13}$$

where $\widetilde{D}^e$ is an arbitrary composition of the Lie derivatives $L_{\widetilde{f}^e}$ and $L_{\widetilde{g}_j^e}$, $j = 1, \ldots, \nu+1$. Now, fix $j \in \{1, \ldots, \nu\}$. First, we will prove (2.10) with $\mu = \nu$. Assume that $\beta_{ij}(x^e) \neq 0$ in a neighbourhood of x_0 for some $i \neq \{j, \nu+1$. We will show that this leads to a contradiction. For, let $\widetilde{g}_0^e = \widetilde{f}^e$. Moreover, let us define recursively the following vector fields of $I\!R^{n+n^w}$ (depending on j)

$$X_{l_0}^e(x,w) = \sum_{i\neq j} \bar{g}_i(x)\beta_{ij}(x,w) + \bar{X}_{l_0}^e(x,w),$$

$$X_{l_s\ldots l_0}^e(x,w) = [\widetilde{g}_{l_s}^e, X_{l_{s-1}\ldots l_0}^e](x,w) + \sum_{i\neq j} \bar{g}_i(x)\varphi_{l_s\ldots l_0 i}(x,w) + \bar{X}_{l_s\ldots l_0}^e(x,w),$$

$$s \geq 1,\ l_s \in \{0, \ldots, \nu+1\}, \tag{2.14}$$

where $\bar{X}_{l_0}^e, \ldots, \bar{X}_{l_s\ldots l_0}^e \in \mathcal{G}^w$ and the smooth (possibly vector–valued) functions $\varphi_{l_s\ldots l_0 i}(x,w)$ are chosen in such a way to satisfy the following equalities

$$dh_i^e(x,w)(\widetilde{g}_j^e(x,w) - \sum_{i\neq j} \bar{g}_i(x)\beta_{ij}(x,w) - \bar{X}_{l_0}^e(x,w)) = 0,$$

$$dh_i^e(x,w)([\widetilde{g}_{l_s}^e, [\widetilde{g}_{l_{s-1}}^e, [\ldots, [\widetilde{g}_{l_1}^e, \widetilde{g}_j^e]\ldots]](x,w) - [\widetilde{g}_{l_s}^e, X_{l_{s-1}\ldots l_0}^e](x,w) - \sum_{i\neq j} \bar{g}_i(x)\varphi_{l_s\ldots l_0 i}(x,w) +$$
$$- \bar{X}_{l_s\ldots l_0}^e(x,w)) = 0,$$

for $i \neq j$. This choice is always possible since from (2.13) and straightforward computations we obtain

$$[\widetilde{g}_{l_s}^e, [\widetilde{g}_{l_{s-1}}^e, [\ldots, [\widetilde{g}_{l_1}^e, \widetilde{g}_j^e]\ldots]](x,w) = [\widetilde{g}_{l_s}^e, X_{l_{s-1}\ldots l_0}^e](x,w) + G(x)\Phi_{l_s\ldots l_0}(x,w) + \bar{X}_{l_s\ldots l_0}^e(x,w)$$

for some $m \times 1$ vectors $\Phi_{l_s\ldots l_0}(x,w)$, with smooth entries, and smooth vector fields $\bar{X}_{l_s\ldots l_0}^e \in \mathcal{G}^w$.

In what follows, for simplicity of notation, if not otherwise stated, we will use x^e instead of (x,w). Let $\bar{\Delta}^e$ (depending on j) be the smooth distribution spanned by the vector fields (2.14). On an open and dense subset $U^e \subset \mathbb{R}^{n+n_w}$ the distribution $\bar{\Delta}^e$ has constant dimension. This, together with (2.14), implies that

$$[\widetilde{g}_j^e, \bar{\Delta}^e] \subset \bar{\Delta}^e + \mathcal{G}^e, \qquad j = 0, \ldots, \nu + 1$$

for all $x^e \in U^e$ and for $j = 1, \ldots, \nu + 1$. Moreover, from (2.13), it follows that $\bar{\Delta}^e \subset \bigcap_{i \neq j} \mathcal{K}_i^e$. Now, denote by $\Delta^e(\bigcap_{i \neq j} \mathcal{K}_i^e)$ the family of smooth distributions $\Delta^e \subset \mathbb{R}^{n+n^w}$, which are contained in $\bigcap_{i \neq j} \mathcal{K}_i^e$ and satisfy

$$[\widetilde{g}_j^e, \Delta^e] \subset \Delta^e + \mathcal{G}^e, \qquad j = 0, \ldots, \nu + 1. \tag{2.15}$$

Clearly, $\Delta^e(\bigcap_{i \neq j} \mathcal{K}_j)$ is closed under the sum of distributions. Thus, there exists $\Delta^{e*} \in \Delta^e(\bigcap_{i \neq j} \mathcal{K}_i^e)$, which is maximal in the sense that any other distribution Δ^e, contained in $\bigcap_{i \neq j} \mathcal{K}_i^e$ and satisfying (2.15), is contained in Δ^{e*}. Note that by construction $\bar{\Delta}^e \subset \Delta^{e*}$ on U^e.

Let $\mathcal{V}_j^*$ be the maximal controlled invariant distribution for $\widetilde{\Sigma}^*$, contained in $\bigcap_{i \neq j} \mathcal{K}_j$. Suppose for a moment that $\Delta^{e*} = \mathcal{V}_j^* + \mathcal{G}^w$ for *all* x^e in a neighbourhood of x_0^e (here, $\mathcal{V}_j^*$ is thought of as a distribution of $\mathbb{R}^{n+n_w}$). Since by assumption $\mathcal{V}_j^* + \mathcal{G}^w$ has constant dimension for all x^e in a neighbourhood of x_0^e, it follows from lemma **2**.3.4 of [55] that

$$\bar{\Delta}^e \subset \mathcal{V}_j^* + \mathcal{G}^w \tag{2.16}$$

for *all* x^e in a neighbourhood of x_0^e. Now, we claim that the distribution $(\sum_{i \neq j} \mathcal{V}_i^*) \cap \mathcal{V}_j^*$ is controlled invariant for $\widetilde{\Sigma}^*$ and is contained in $\bigcap_{i=1}^{\nu} \mathcal{K}_i$. Indeed, since $\sum_{i \neq j} \mathcal{V}_i^* \subset \mathcal{K}_j$, it immediately follows that $(\sum_{i \neq j} \mathcal{V}_i^*) \cap \mathcal{V}_j^* \subset \bigcap_{i=1}^{\nu} \mathcal{K}_i$. Since $\widetilde{\Sigma}^*$ has vector relative degree at x_0^e, from lemma **5**.1.2 and remark **6**.3.15 of [55] it follows that $\sum_{i \neq j} \mathcal{V}_i^*$ is controlled invariant for $\widetilde{\Sigma}^*$. We have

$$\begin{aligned}[\widetilde{g}_i, (\sum_{i \neq j} \mathcal{V}_i^*) \cap \mathcal{V}_j^*] &\subset (\sum_{i \neq j} \mathcal{V}_i^* + \mathcal{G}) \cap (\mathcal{V}_j^* + \mathcal{G}) = (\sum_{i \neq j} \mathcal{V}_i^* + \mathcal{V}_j^* \cap \mathcal{G}) \cap \mathcal{V}_j^* + \mathcal{G} = \\ &= (\sum_{i \neq j} \mathcal{V}_i^*) \cap \mathcal{V}_j^* + \mathcal{G}\end{aligned}$$

for $i = 1, \ldots, \nu + 1$. This proves our claim.

Now, note that $X_{l_0}^e \in \sum_{i \neq j} \mathcal{V}_i^* + \mathcal{G}^w$. This, together with (2.16), implies that

$$X_{l_0}^e \subset (\sum_{i \neq j} \mathcal{V}_i^*) \cap \mathcal{V}_j^* \cap \mathcal{G} + \mathcal{G}^w. \tag{2.17}$$

It is not hard to show that the maximal controllability distribution for $\widetilde{\Sigma}^*$, contained in $\bigcap_{i=1}^{\nu} \mathcal{K}_i$, is equal to $\mathcal{Q}^*$. Moreover, let $\mathcal{V}^*$ be the maximal controlled invariant distribution for $\widetilde{\Sigma}^*$, contained in $\bigcap_{i=1}^{\nu} \mathcal{K}_i$. Since $\sum_{i \neq j} \mathcal{V}_i^* \cap \mathcal{V}_j^*$ is controlled invariant and it is contained in $\bigcap_{i=1}^{\nu} \mathcal{K}_i$, it follows that $\sum_{i \neq j} \mathcal{V}_i^* \cap \mathcal{V}_j^* \subset \mathcal{V}^*$. From (2.17) we conclude that $X_{l_0}^e \in \mathcal{Q}^* \cap \mathcal{G} + \mathcal{G}^w$ (here, $\mathcal{Q}^* \cap \mathcal{G}$ is thought of as a distribution of $\mathbb{R}^{n+n_w}$). This gives the claimed contradiction, since $X_{l_0}^e = \sum_{i \neq j} \widetilde{g}_i \beta_{ij} + \bar{X}_{l_0}^e$ and $\operatorname{span}\{\widetilde{g}_1, \ldots, \widetilde{g}_\nu\} \cap \mathcal{Q}^* = 0$ for all x in a neighbourhood of x_0. In a similar way, we can prove that $\beta_{i,\nu+1}(x^e) = 0$ for all x^e in a neighbourhood of x_0^e and for $i \neq \nu + 1$. In what follows, we will assume that $\beta_{ij}(x^e) = 0$ for all x^e in a neighbourhood of x_0^e, for $j = 1, \ldots, \nu + 1$ and $i \neq \{j, \nu + 1\}$ (in this case we will consider the vector fields

$$X_{l_0}^e(x, w) = \sum_{i=1}^{\nu+1} \widetilde{g}_i(x) \beta_{ij}(x, w) + \bar{X}_{l_0}^e(x, w),$$

$$X_{l_s \ldots l_0}^e(x, w) = [\widetilde{g}_{l_s}^e, X_{l_{s-1} \ldots l_0}^e](x, w) + \sum_{i=1}^{\nu+1} \widetilde{g}_i(x) \varphi_{l_s \ldots l_0 i}(x, w) + \bar{X}_{l_s \ldots l_0}^e(x, w),$$

$$s \geq 1,\ l_s \in \{0, \ldots, \nu + 1\}).$$

Finally, we will prove (2.11) and (2.12) with $\mu = \nu$ by using induction. Fix $j \in \{1, \ldots, \nu\}$. First, note that, since $\beta_{ij}(x^e) = 0$ for $i \neq \{j, \nu + 1\}$ and for all x^e in a neighbourhood of x_0^e, we have

$$0 = dh_i^e(x, w)([\widetilde{f}^e, \widetilde{g}_j^e])(x, w) = dh_i(x)\Big(\sum_{i \neq j} \widetilde{g}_i(x)(-L_{\widetilde{g}_j^e} \alpha_i(x, w))\Big), \qquad i \neq j,$$

$$0 = dh_i^e\Big(\sum_{i \neq j} [\widetilde{g}_i^e, \widetilde{g}_j^e]\Big)(x, w) = dh_i(x)(\widetilde{g}_i(x)(-L_{\widetilde{g}_j^e} \beta_{ii}(x, w))), \qquad i \neq j.$$

Reasoning as above with $\beta_{ij}(x^e)$ replaced by $L_{\widetilde{g}_j^e} \alpha_i(x^e)$ and $L_{\widetilde{g}_j^e} \beta_{ii}(x^e)$, respectively, we can prove that

$$\begin{aligned} L_{\widetilde{g}_j^e} \alpha_i(x^e) &= 0, \qquad i \neq j, \\ L_{\widetilde{g}_j^e} \beta_{ii}(x^e) &= 0, \qquad i \neq j \end{aligned} \tag{2.18}$$

for all x^e in a neighbourhood of x_0^e. Now, assume that for all x^e in a neighbourhood of x_0^e

$$\begin{aligned} L_{\widetilde{g}_j^e} \widetilde{D}^{ek} \alpha_i(x^e) &= 0, \qquad i \neq j, \\ L_{\widetilde{g}_j^e} \widetilde{D}^{ek} \beta_{ii}(x^e) &= 0, \qquad i \neq j, \end{aligned} \tag{2.19}$$

where $\widetilde{D}^{ek}$ is an arbitrary composition of *at most* $k \geq 1$ Lie derivatives $L_{\widetilde{f}^e}$ and $L_{\widetilde{g}_i^e}$,

$i = 1, \ldots, \nu + 1$. We have

$$0 = dh_i^e(x,w)([\widetilde{f}^e, [\widetilde{g}^e_{l_{k+2}}, [\ldots, [\widetilde{g}^e_{l_1}, \widetilde{g}^e_j] \ldots]]])(x,w) =$$
$$= dh_i(x)(\widetilde{g}_i(x)(-L_{[\widetilde{g}^e_{l_{k+2}},[\ldots,[\widetilde{g}^e_{l_1},\widetilde{g}^e_j]\ldots]]}\alpha_i(x,w)), \qquad j \neq i,$$
$$0 = dh_i^e(x,w)([\widetilde{g}^e_i, [\widetilde{g}^e_{l_{k+2}}, [\ldots, [\widetilde{g}^e_{l_1}, \widetilde{g}^e_j] \ldots]]])(x,w) =$$
$$= dh_i(x)(\widetilde{g}_i(x)(-L_{[\widetilde{g}_{l_{k+2}},[\ldots,[\widetilde{g}^e_{l_1},\widetilde{g}^e_j]\ldots]]}\beta_{ii}(x,w)), \qquad i \neq j.$$

Reasoning as above with $\beta_{ij}(x^e)$ replaced by $L_{[\widetilde{g}_{l_{k+2}},[\ldots,[\widetilde{g}^e_{l_1},\widetilde{g}^e_j]\ldots]]}\alpha_i(x^e)$ and $L_{[\widetilde{g}_{l_{k+2}},[\ldots,[\widetilde{g}^e_{l_1},\widetilde{g}^e_j]\ldots]]}$ $\beta_{ii}(x^e)$, respectively, we can prove that

$$L_{[\widetilde{g}_{l_{k+2}},[\ldots,[\widetilde{g}^e_{l_1},\widetilde{g}^e_j]\ldots]]]}\alpha_i(x^e) = 0, \qquad i \neq j,$$
$$L_{[\widetilde{g}_{l_{k+2}},[\ldots,[\widetilde{g}^e_{l_1},\widetilde{g}^e_j]\ldots]]]}\beta_{ii}(x^e) = 0, \qquad i \neq j$$

for all x^e in a neighbourhood of x_0^e. This, together with (2.18) and the induction hypothesis (2.19), implies that (2.19) holds with k replaced by $k+1$.

The proof of the lemma is complete if we prove that $\Delta^{e*} = \mathcal{V}_j^* + \mathcal{G}^w$ for all x^e in a neighbourhood of x_0^e.

For, let us consider the following sequence of codistributions (depending on j)

$$\begin{aligned} \Omega_0^e &= \sum_{i \neq j} \operatorname{span}\{dh_i^e\}, \\ \Omega_k^e &= L_{\widetilde{f}^e}(\Omega_{k-1}^e \cap (\mathcal{G}^e)^\perp) + \sum_{i=1}^{\nu+1} L_{\widetilde{g}_i^e}(\Omega_{k-1}^e \cap (\mathcal{G}^e)^\perp) + \Omega_{k-1}^e. \end{aligned} \tag{2.20}$$

Using essentially the same arguments of lemma 6.3.2 of [55], it is not hard to show that, if there exists $k^* \geq 0$ such that $\Omega_{k^*}^e = \Omega_{k^*+1}^e$ and Ω_k^e and $\Omega_k^e \cap (\mathcal{G}^e)^\perp$ have constant dimension for all x^e in a neighbourhood of x_0^e and for all k, we have $\ker\{\Omega_{k^*}^e\} = \Delta^{e*}$. Moreover, if $\Omega_k^e = \Omega_{k+1}^e$ for some $k \geq 0$, then $\Omega_k^e = \Omega_h^e$ for all $h \geq k$.

To prove the lemma, we will show that there exists $k^* \geq 0$ such that $\ker\{\Omega_{k^*}^e\} = \ker\{\Omega_{k^*+1}^e\} = \mathcal{V}_j^* + \mathcal{G}^w$ (here, $\mathcal{V}_j^*$ is thought of as a distribution of $\mathbb{R}^{n+n^w}$) and that Ω_k^e and $\Omega_k^e \cap (\mathcal{G}^e)^\perp$ have constant dimension for all x^e in a neighbourhood of x_0 and for all k. For, by assumption Ω_0^e has constant dimension in a neighbourhood of x_0^e. The codistribution $\Omega_0^e \cap (\mathcal{G}^e)^\perp$ is spanned at each x^e by the rows of $\Omega_0^e(x^e)$ which annihilate the vectors of $\mathcal{G}^e(x^e)$. Since $\widetilde{\Sigma}^*$ has vector relative degree at x_0^e, $\Omega_0^e \cap (\mathcal{G}^e)^\perp$ is spanned by the covector fields $dh_{i_l}^e$, $i_l \in \{1, \ldots, j-1, j+1, \ldots, \nu\}$, such that $\langle dh_{i_l}^e, G^e\rangle(x^e) = 0$. Since $L_{\widetilde{f}^e} dh_{i_l}^e = dL_{\widetilde{f}^e} h_{i_l}^e$ and $L_{\widetilde{g}_i^e} dh_{i_l}^e = dL_{\widetilde{g}_i^e} h_{i_l}^e$, it follows that $L_{\widetilde{f}^e}(\Omega_0^e \cap (\mathcal{G}^e)^\perp) = L_{f^e}(\Omega_0^e \cap (\mathcal{G}^e)^\perp)$ and

$L_{\widetilde{g}_i^e}(\Omega_0^e \cap (\mathcal{G}^e)^\perp) = 0$ for $i = 1, \ldots, \nu+1$. Suppose now that for some $k > 0$ the codistribution Ω_k^e has constant dimension in a neighbourhood of x_0^e and that $\Omega_k^e \cap (\mathcal{G}^e)^\perp$ is spanned by the covector fields $dL_{f^e}^{k_{i_l}} h_{i_l}^e$, $i_l \in \{1, \ldots, j-1, j+1, \ldots, \nu\}$, such that $\langle dL_{f^e}^{k_{i_l}} h_{i_l}^e, G^e \rangle (x^e) = 0$ for all k_{i_l} such that $0 \leq k_{i_l} \leq \min\{k, r_{i_l} - 2\}$ ($r_1, \ldots, r_\nu$ is the vector relative degree of $\widetilde{\Sigma}^*$). Similarly, we have $L_{\widetilde{f}^e}(\Omega_k^e \cap (\mathcal{G}^e)^\perp) = L_{f^e}(\widetilde{\Omega}_k^e \cap (\mathcal{G}^e)^\perp)$ and $L_{\widetilde{g}_i^e}(\Omega_k^e \cap (\mathcal{G}^e)^\perp) = 0$ for $i = 1, \ldots, \nu + 1$. Since $\widetilde{\Sigma}^*$ has vector relative degree at x_0^e, the codistribution $\widetilde{\Omega}_{k+1}^e$ has always constant dimension in a neighbourhood of x_0^e (see lemma **5**.12 of [55] for a proof). Clearly $\Omega_k^e \cap (\mathcal{G}^e)^\perp$ is spanned by the covector fields $dL_{f^e}^{k_{i_l}} h_{i_l}^e$, $i_l \in \{1, \ldots, j-1, j+1, \ldots, \nu\}$, such that $\langle dL_{f^e}^{k_{i_l}} h_{i_l}^e, G^e \rangle (x^e) = 0$ for all k_{i_l} such that $0 \leq k_{i_l} \leq \min\{k+1, r_{j_l}\} - 1$. The sequence (2.20) stops for $k = k^* = \max_i \{r_i - 2\}$, when $\Omega_{k^*}^e \cap (\mathcal{G}^e)^\perp = \Omega_{k^*+1}^e \cap (\mathcal{G}^e)^\perp$. This, together with the fact that $(\mathcal{V}_j^*)^\perp = \mathrm{span}\{dh_i, \ldots, dL_{\widetilde{f}}^{r_i - 1} h_i : i \neq j\}$ (corollary **6.3.14** of [55]), proves our thesis. □

Now, we wish to generalize lemma **3.4.4** and remark **3.4.5** to the class of systems here considered. Toward this end, we need some preliminary results. First, let us consider an invertible feedback law which solves ND for $\widetilde{\Sigma}^*$ and

$$(\alpha_1^T(x,w) \;\; \cdots \;\; \alpha_{\nu+1}^T(x,w))^T, \; (\beta_1(x,w) \;\; \cdots \;\; \beta_{\nu+1}(x,w))^T,$$

$$(\gamma_1^T(x,w) \;\; \cdots \;\; \gamma_{\nu+1}^T(x,w))^T, \; (\delta_1(x,w) \;\; \cdots \;\; \delta_{\nu+1}(x,w))^T$$

be the partitions of $\alpha(x,w)$, $\beta(x,w)$, $\gamma(x,w)$ and $\delta(x,w)$ corresponding to the partition $u_1, \ldots, u_{\nu+1}$ of the input vector u in $\widetilde{\Sigma}^*$. By definition **3**.1.5, the system

$$\begin{aligned} u &= \alpha(x,w) + \beta(x,w)v, \\ \dot{x} &= f(x) + G(x)\alpha(x,w) + G(x)\beta(x,w)v, \\ \dot{w} &= \gamma(x,w) + \delta(x,w)v, \end{aligned}$$

with $\dot{x} = f(x) + G(x)v$ being $\widetilde{\Sigma}^*$, is invertible in the sense of Singh. Now, let dim $((\mathrm{span}\,\{\widetilde{g}_i\})(x)) = m_i$ and dim $((\mathrm{span}\,\{\widetilde{g}_i^e\})(x^e)) = m_i^e$. From theorem 1 of [76] it follows that there exists a dynamic feedback law, defined on an open neighbourhood of the origin of the state space, such that the system, resulting from plugging this feedback law into (**3**.1.5), is noninteractive. Now, if $m_i = 1$, we can proceed as in the proof of proposition **3**.1.3 ([86]) and conclude that each system

$$\begin{aligned} u_i &= \alpha_i(x,w) + \beta_{ii}(x,w)v_i, \\ \dot{x} &= f(x) + G(x)\alpha(x,w) + G(x)\beta(x,w)v, \\ \dot{w} &= \gamma(x,w) + \delta(x,w)v, \end{aligned} \tag{2.21}$$

with $i = 1, \ldots, \nu$, has relative degree at x_0^e (if we consider u_i as output vector). On the other hand, if $m_i > 1$, for each system (2.21) (with input vector v and output vector u_i) there exists a dynamic feedback law, defined on an open neighbourhood of the origin of the state space, such that the system, resulting from plugging this feedback law into (2.21), is noninteractive (with each input controlling one input). Let us consider the following auxiliary system (depending on i)

$$\begin{aligned} \dot{\zeta} &= \alpha_i(x, w) + \beta_{ii}(x, w) v_i, \\ \dot{x} &= f(x) + G(x)\alpha(x, w) + G(x)\beta(x, w) v, \\ \dot{w} &= \gamma(x, w) + \delta(x, w) v, \\ \bar{h}(\zeta, x, w) &= \zeta, \end{aligned} \tag{2.22}$$

where $\bar{h}(\zeta, x, w)$ is a vector of fictitious outputs and let $(\zeta^T, x^T, w^T)^T = \bar{x}^e$, with $\bar{x}_0^e = 0$. Moroever, let

$$\bar{f}^e(\bar{x}^e) = \begin{pmatrix} \alpha_i(x, w) \\ \tilde{f}^e(x, w) \end{pmatrix}, \ \bar{g}_j^e(\bar{x}^e) = \begin{cases} \begin{pmatrix} \beta_{ii}(x, w) \\ \tilde{g}_i^e(x, w) \end{pmatrix} & \text{if } j = i \\ \begin{pmatrix} 0_{m_i \times m_j^e} \\ \tilde{g}_j^e(x, w) \end{pmatrix} & \text{else}, \end{cases}$$

$$\bar{G}^e(\bar{x}^e) = (\bar{g}_1^e(\bar{x}^e) \quad \cdots \quad \bar{g}_{\nu+1}^e(\bar{x}^e)), \ \bar{\mathcal{G}}^e = \mathrm{span}\{\bar{G}^e\}.$$

With the above notations, we introduce the following sequence of distributions

$$\begin{aligned} \mathcal{S}_0^e &= \bar{\mathcal{G}}^e, \\ \mathcal{S}_k^e &= [\bar{f}^e, \ker\{d\zeta\} \cap \mathcal{S}_{k-1}^e] + \sum_{j=1}^{\nu+1} [\bar{g}_j^e, \ker\{d\zeta\} \cap \mathcal{S}_{k-1}^e] + \mathcal{S}_{k-1}^e. \end{aligned} \tag{2.23}$$

Note that if the distributions $\mathcal{S}_k^e$ and $\mathcal{S}_k^e \cap \ker\{d\zeta\}$ have constant dimension for all k, then there exists $k^* \geq 0$ such that $\mathcal{S}_{k^*}^e = \mathcal{S}_k^e$ for all $k \geq k^*$. Moreover, $\mathcal{S}^{e*} = \mathcal{S}_{k^*}^e$ is the *minimal* distribution which contains $\bar{\mathcal{G}}^e$ and satisfies $[\bar{f}^e, \mathcal{S}^{e*} \cap \ker\{d\zeta\}] \subset \mathcal{S}^{e*}$ and $[\bar{g}_j^e, \mathcal{S}^{e*} \cap \ker\{d\zeta\}] \subset \mathcal{S}^{e*}$, $j = 1, \ldots, \nu+1$ (see also [66]).

We proceed now to show how the algorithm (2.23) can be implemented in practice.

To compute $\mathcal{S}_1^e$, we have to compute first $\bar{\mathcal{G}}^e \cap \ker\{d\zeta\}$. Let $A_0(x^e) = \beta_{ii}(x^e)$. Moreover, assume that $A_0(x^e)$ has constant rank s_0 in a neighbourhood of x_0^e and $\bar{G}^e(\bar{x}^e)$ has constant rank in a neighbourhood of $\bar{x}_0^e$ (or, equivalently, that $\mathcal{S}_0^e \cap \ker\{d\zeta\}$ and $\mathcal{S}_0^e$ have constant dimension in a neighbourhood of $\bar{x}_0^e$). Moreover, let l_0 be the number of columns of $A_0(x^e)$. It is always possible to find matrices $A_{0h}(x^e)$ and $B_{0h}(x^e)$, $h = 1, 2$, with smooth entries and such that (after possibly permuting the columns of $A_0(x^e)$) we have

$$A_0(x^e) = (\, A_{01}(x^e) \quad A_{02}(x^e)\,)\ ,\ A_0B_{01}(x^e) = A_{01}(x^e)\ ,\ A_0B_{02}(x^e) = 0_{m_i\times(l_0-s_0)},$$

with $(\, B_{01}(x^e) \quad B_{02}(x^e)\,)$ invertible for all x^e in a neighbourhood of x_0^e . Moreover, using (2.10), (2.11) and (2.12), it is not hard to show that the matrices $B_{0h}(x^e)$, $h = 1, 2$, can be chosen in such a way that

$$L_{\widetilde{g}_j^e}\widetilde{D}^e B_{0h}(x^e) = 0, \qquad j \neq i, \tag{2.24}$$

where $\widetilde{D}^e$ is an arbitrary composition of the Lie derivatives $L_{\widetilde{f}^e}$ and $L_{\widetilde{g}_j^e}$, $j = 1, \ldots, \nu + 1$. Let us define the following vector fields

$$\begin{aligned}
\bar{T}_{01} &= (\, \bar{g}_1^e \quad \cdots \quad \bar{g}_{i-1}^e \quad \bar{g}_{i+1}^e \quad \cdots \quad \bar{g}_{\nu+1}^e\,),\\
\bar{T}_{02} &= \bar{g}_i^e B_{02},\\
T_{02}^{(i)} &= (\, \bar{T}_{01} \quad \bar{T}_{02}\,),\\
T_{01} &= \bar{g}_i^e B_{01},\\
T_1 &= (\, T_{01} \quad T_{02} \quad [\bar{g}_1^e, T_{02}] \quad \cdots \quad [\bar{g}_{\nu+1}^e, T_{02}] \quad [\bar{f}^e, T_{02}]\,),
\end{aligned}$$

and

$$\begin{aligned}
\bar{R}_{02} &= \widetilde{g}_i^e B_{02},\\
R_{01} &= \widetilde{g}_i^e B_{01}.
\end{aligned}$$

The distribution $\bar{\mathcal{G}}^e \cap \ker\{d\zeta\}$ is spanned by the vector fields T_{02} and the distribution $\mathcal{S}_1^e$ is spanned by the vector fields T_1.

At the step $(k+1)$–th, $k \geq 1$, to compute $\mathcal{S}_{k+1}^e$ we have to compute first $\mathcal{S}_k^e \cap \ker\{d\zeta\}$. Let

$$A_k(x^e) = (\, A_{k-1,1}(x^e) \quad L_{\bar{R}_{k-1,2}}\beta_{ii}(x^e) \quad L_{\bar{R}_{k-1,2}}\alpha_i(x^e)\,).$$

Moreover, assume that $A_k(x^e)$ has constant rank s_k in a neighbourhood of x_0^e and $T_k(\bar{x}^e)$ has constant rank in a neighbourhood of $\bar{x}_0^e$ (or, equivalently, that $\mathcal{S}_k^e \cap \ker\{d\zeta\}$ and $\mathcal{S}_k^e$ have constant dimension in a neighbourhood of $\bar{x}_0^e$). In this case, it is possible to find matrices $A_{kh}(x^e)$ and $B_{kh}(x^e)$, $h = 1, 2$, with smooth entries and such that (after possibly permuting the columns of $A_k(x^e)$) we have

$$A_k(x^e) = (\, A_{k1}(x^e) \quad A_{k2}(x^e)\,)\ ,\ A_kB_{k1}(x^e) = A_{k1}(x^e)\ ,\ A_kB_{k2}(x^e) = 0_{m_i\times(l_k-s_k)}$$

and $(B_{k1}(x^e) \quad B_{k2}(x^e))$ is invertible for all x^e in a neighbourhood of x_0^e. Moreover, using (2.10), (2.11) and (2.12), it is not hard to show that the matrices $B_{kh}(x^e)$, $h = 1,2$, can be chosen in such a way that

$$L_{\widetilde{g}_j^e} \widetilde{D}^e B_{kh}(x^e) = 0, \qquad j \neq i, \tag{2.25}$$

where $\widetilde{D}^e$ is an arbitrary composition of the Lie derivatives $L_{\widetilde{f}^e}$ and $L_{\widetilde{g}_j^e}$, $j = 1, \dots, \nu+1$. Let us define the following vector fields

$$\begin{aligned}
\bar{T}_{k1} &= (T_{k-1,2} \quad [\bar{g}_1^e, T_{k-1,2}] \quad \cdots \quad [\bar{g}_{i-1}^e, T_{k-1,2}] \quad [\bar{g}_i^e, \bar{T}_{k-1,1}] \quad [\bar{g}_{i+1}^e, T_{k-1,2}] \quad \cdots \\
&\qquad \cdots \quad [\bar{g}_{\nu+1}^e, T_{k-1,2}] \quad [\bar{f}^e, \bar{T}_{k-1,1}]), \\
\bar{T}_{k2} &= (T_{k-1,1} \quad [\bar{g}_i^e, \bar{T}_{k-1,2}] \quad [\bar{f}^e, \bar{T}_{k-1,2}]) B_{k2}, \\
T_{k2} &= (\bar{T}_{k1} \quad \bar{T}_{k2}), \\
T_{k1} &= (T_{k-1,1} \quad [\bar{g}_i^e, \bar{T}_{k-1,2}] \quad [\bar{f}^e, \bar{T}_{k-1,2}]) B_{k1}, \\
T_{k+1} &= (T_{k1} \quad T_{k2} \quad [\bar{g}_1^e, T_{k2}] \quad \cdots \quad [\bar{g}_{\nu+1}^e, T_{k2}] \quad [\bar{f}^e, T_{k2}]),
\end{aligned}$$

and

$$\begin{aligned}
\bar{R}_{k2} &= \left(R_{k-1,1} \quad [\widetilde{g}_i^e, \bar{R}_{k-1,2}] \quad [\widetilde{f}^e, \bar{R}_{k-1,2}]\right) B_{k2}, \\
R_{k1} &= \left(R_{k-1,1} \quad [\widetilde{g}_i^e, \bar{R}_{k-1,2}] \quad [\widetilde{f}^e, \bar{R}_{k-1,2}]\right) B_{k1}.
\end{aligned}$$

The distribution $\mathcal{S}_k^e \cap \ker\{d\zeta\}$ is spanned by the vector fields T_{k2}, since, as a direct consequence of (2.10), (2.11), (2.12), (2.24) and (2.25), the only vector fields, which have the first m_i components possibly not identically equal to zero, are of the form T_{h1}, $[\bar{g}_i^e, \bar{T}_{h2}]$ or $[\bar{f}^e, \bar{T}_{h2}]$, with $h \leq k-1$, and, moreover, the first m_i rows of $(T_{h1} \quad [\bar{g}_i^e, \bar{T}_{h2}] \quad [\bar{f}^e, \bar{T}_{h2}])(\bar{x}^e)$ are identically equal (up to sign) to $A_{h+1}(x^e)$ for $h \leq k-1$. On the other hand, the distribution $\mathcal{S}_{k+1}^e$ is spanned by the vector fields T_{k+1}. This completes the description of a possible implementation of the algorithm (2.23).

In what follows, we will assume that the distributions $\mathcal{S}_k$ and $\mathcal{S}_k \cap \ker\{d\zeta\}$, $k \geq 0$, have constant dimension in a neighbourhood of $\bar{x}_0^e$. In this case there exists indeed $k^* \geq 0$ such that $\mathcal{S}_{k^*}^e = \mathcal{S}_{k^*+1}^e = \mathcal{S}^{e*}$ and we will say that $\mathcal{S}^{e*}$ is *regularly computable* at $\bar{x}_0^e$. The fact that (2.22) can be rendered noninteractive via dynamic feedback law has an interesting consequence in terms of $\mathcal{S}^{e*}$. This is shown in the next proposition.

Proposition 2.8. Assume that $\mathcal{S}^{e*}$ is regularly computable at $\bar{x}_0^e$. Then, under our assumptions, $\mathcal{S}^{e*} + \ker\{d\zeta\} = \mathbb{R}^{m_i+n+n^w}$ for all $\bar{x}^e$ in a neighbourhood of $\bar{x}_0^e$. □

Proof. First, assume that $m_i = 1$. We know that, in this case, (2.22) has relative degree at $\bar{x}_0^e$. Without loss of generality, we can suppose that (2.22) is already in *normal* form (see

[55], pp. 152). It is straightforward to check that in this case $\mathcal{S}^{e*} + \ker\{d\zeta\} = \mathbb{R}^{m_i+n+n^w}$ for all $\bar{x}^e$ in a neighbourhood of $\bar{x}^e_0$.

Now, assume that $m_i > 1$. Following [76], from our assumptions it follows that there exists a feedback law (depending on i)

$$\begin{aligned} v_i &= \alpha(\bar{x}^e, \eta) + \beta(\bar{x}^e, \eta)v', \\ \dot{\eta} &= \gamma(\bar{x}^e, \eta) + \delta(\bar{x}^e, \eta)v', \end{aligned} \tag{2.26}$$

such that the input–output behaviour of the system (2.22)–(2.26) is described by

$$(\zeta_k)^{(l_k)} = v'_{j_k}, \qquad k = 1, \ldots, m_i, \tag{2.27}$$

where the subscript (l_k) denotes the number of derivatives with respect to time, ζ_k is the k–th component of the output vector ζ, v'_{j_k} is the j_k–th component of the input vector v', $l_k \geq 0$ and $j_k \in \{1, \ldots, m\}$ with $j_r \neq j_s$ for $r, s = 1, \ldots, m_i$ and $r \neq s$. Let

$$\widehat{f}^e = \begin{pmatrix} \bar{f}^e(\bar{x}^e) + \bar{G}^e(\bar{x}^e)\alpha(\bar{x}^e, \eta) \\ \gamma(\bar{x}^e, \eta) \end{pmatrix},$$

$$\widehat{G}^e(\bar{x}^e, \eta) = \begin{pmatrix} \bar{G}^e(\bar{x}^e)\beta(\bar{x}^e, \eta) \\ \delta(\bar{x}^e, \eta) \end{pmatrix}, \ \widehat{\mathcal{G}}^e = \operatorname{span}\{\widehat{G}^e\},$$

and $\widehat{g}^e_j(\bar{x}^e, \eta)$ be the j–th column (block) of $\widehat{G}^e(\bar{x}^e, \eta)$. From (2.27), it follows that, after possibly changing coordinates, (2.22)–(2.27) is in the form

$$\begin{aligned} \dot{\theta}_{k1} &= \theta_{k2}, \\ \cdots &= \cdots \\ \dot{\theta}_{kl_k} &= v'_{j_k}, \qquad k = 1, \ldots, m_i, \\ \dot{\varphi} &= \psi(\theta_{11}, \ldots, \theta_{1l_1}, \ldots, \theta_{m_i 1}, \ldots, \theta_{m_i l_{m_i}}, \varphi) + \\ &\quad + \lambda(\theta_{11}, \ldots, \theta_{1l_1}, \ldots, \theta_{m_i 1}, \ldots, \theta_{m_i l_{m_i}}, \varphi)v', \end{aligned}$$

for some (possibly vector–valued) smooth functions ψ and λ and with output vector $\widehat{h}(\theta_{11}, \ldots, \theta_{1l_1}, \ldots, \theta_{m_i 1}, \ldots, \theta_{m_i l_{m_i}}, \varphi) = (\theta_{11}, \ldots, \theta_{m_i 1})^T$. Now, denote by $\widehat{\mathcal{S}}^{e*}$ the minimal distribution which contains $\widehat{\mathcal{G}}^e$ and is such that $[\widehat{f}^e, \widehat{\mathcal{S}}^{e*} \cap \ker\{d\widehat{h}\}] \subset \widehat{\mathcal{S}}^{e*}$ and $[\widehat{g}^e_j, \widehat{\mathcal{S}}^{e*} \cap \ker\{d\widehat{h}\}] \subset \widehat{\mathcal{S}}^{e*}$ for $j = 1, \ldots, \nu + 1$, and assume for simplicity that $\widehat{\mathcal{S}}^{e*}$ is regularly computable at the origin of the state space (the proof of the lemma can be repeated also after removing this assumption). By straightforward computations, it can be seen that the distribution $\widehat{\mathcal{S}}^{e*}$ is such that

$$\langle d\widehat{h}, \widehat{\mathcal{S}}^{e*} \rangle(\bar{x}^e_0, \eta) = \mathbb{R}^{m_i} \tag{2.28}$$

($\langle d\widehat{h}, \widehat{S}^{e*}\rangle(\bar{x}_0^e, \eta)$ denotes the subspace of $\mathbb{R}^{m_i}$ spanned by the vectors $\langle d\bar{h}, \tau\rangle(\bar{x}_0^e, \eta)$ for $\tau \in \widehat{S}^{e*}$). On the other hand, it is not hard to show by induction that, if π is the natural projection $\pi(\bar{x}^e, \eta) = \bar{x}^e$ and $\pi_{*(\bar{x}^e,\eta)}$ its differential at $(\bar{x}^e, \eta)$, we have $\pi_{*(\bar{x}^e,\eta)}\widehat{S}^{e*}(\bar{x}^e, \eta) \subset S^{e*}(\pi(x_i^e, \eta))$. This, together with (2.28), proves our thesis. □

An interesting consequence of proposition 2.8 is the following generalization of proposition **3**.4.4. Let us consider the distributions $\mathcal{R}_i^e = \langle \widetilde{f}^e, \widetilde{g}_1^e, \ldots, \widetilde{g}_{\nu+1}^e | \operatorname{span}\{\widetilde{g}_i^e, \widetilde{g}_{\nu+1}^e\}\rangle$, $i = 1, \ldots, \nu$, and Δ_{MIX}^e be the distribution defined as in section 2.1, but with $\widetilde{f}$ and $\widetilde{g}_j$, $j = 1, \ldots, \nu+1$, replaced by $\widetilde{f}^e$ and $\widetilde{g}_j$, respectively.

Proposition 2.9. Assume that S^{e*} is regularly computable at $\bar{x}_0^e$ and $S^{e*} + \ker\{d\zeta\} = \mathbb{R}^{m_i+n+n^w}$ for all $\bar{x}^e$ in a neighbourhood of $\bar{x}_0^e$. Moreover, assume that the distributions $\mathcal{R}_i^e$, $i = 1, \ldots, \nu$, are regularly computable at x_0^e and Δ_{MIX}^e has constant dimension in a neighbourhood of x_0^e. Then, for all x^e in a neighbourhood of x_0^e

$$\pi_{*x^e}(\Delta_{MIX}^e + \mathcal{Q}^*)(x^e) = \Delta_{MIX}^*(\pi(x^e)), \tag{2.29}$$

$$\pi_{*x^e}(\mathcal{R}_i^e + \mathcal{Q}^*)(x^e) = \mathcal{R}_i^*(\pi(x^e)), \tag{2.30}$$

where $\pi : \mathbb{R}^{n+n_w} \to \mathbb{R}^n$ is the natural projection $\pi(x^e) = x$ and $\mathcal{Q}^*$ is thought of as a distribution of $\mathbb{R}^{n+n_w}$. □

Proof. Let B_{kh}, $\bar{R}_{k2}$, R_{k1} and s_k be as above and, if $q = \dim(\mathcal{Q}^*(x^e))$, denote by $(A \bmod(\mathcal{Q}^* + \mathcal{G}^w))(x^e)$ the first $n-q$ components (resp. rows) of a given vector (resp. matrix) $A(x^e)$. From (2.10), (2.11), (2.12), (2.24) and (2.25) we obtain the following equalities (up to signs)

$$\begin{aligned}
([\widetilde{f}^e, \bar{R}_{k2}] \bmod(\mathcal{Q}^* + \mathcal{G}^w))(x^e) &= \widetilde{g}_i(x)(L_{\bar{R}_{k2}}\alpha_i)(x^e), \\
([\widetilde{g}_i^e, \bar{R}_{k2}] \bmod(\mathcal{Q}^* + \mathcal{G}^w))(x^e) &= \widetilde{g}_i(x)(L_{\bar{R}_{k2}}\beta_{ii})(x^e), \\
([\widetilde{g}_j^e, \bar{R}_{k2}] \bmod(\mathcal{Q}^* + \mathcal{G}^w))(x^e) &= 0, \qquad j \neq i.
\end{aligned} \tag{2.31}$$

Proposition 2.8 implies that the first m_i rows of $T_{k^*}(\bar{x}^e)$ are linearly independent for all $\bar{x}^e$ in a neighbourhood of $\bar{x}_0^e$. Since the first m_i rows of $(\,T_{k-1,1} \quad [\bar{g}_i^e, \bar{T}_{k-1,2}] \quad [\bar{f}^e, \bar{T}_{k-1,2}]\,)(\bar{x}^e)$ are identically equal (up to sign) to $A_k(x^e)$ for all k, the matrix $A_{k^*1}(x^e)$ has full row rank for all x^e in a neighbourhood of x_0^e and $(R_{k^*1} \bmod (\mathcal{Q}^* + \mathcal{G}^w))(x^e) = \widetilde{g}_i(x)A_{k^*1}(x^e)$. Since $\mathcal{R}_i^e$ is invariant under $\widetilde{f}^e, \widetilde{g}_1^e, \ldots, \widetilde{g}_{\nu+1}^e$ and from (2.24) and (2.25), it follows that $[\widetilde{g}_j^e, \bar{R}_{k2}] \in \mathcal{R}_i^e$, $j = 1, \ldots, \nu+1$, and $[\widetilde{f}^e, \bar{R}_{k2}] \in \mathcal{R}_i^e$. From (2.31) we conclude that $\widetilde{g}_i(\pi(x^e)) \in \pi_{*x^e}(\mathcal{R}_i^e + \mathcal{Q}^*)(x^e)$. By combining induction as in proposition **3**.4.3 and the above facts, we obtain $\pi_{*x^e}(\Delta_{MIX}^e + \mathcal{Q}^*)(x^e) \supset (\Delta_{MIX}^*)(\pi(x^e))$. On the other hand, from

(2.24) and (2.25) it easily follows that $\pi_{*x^e}(\Delta^e_{MIX} + \mathcal{Q}^*)(x^e) \subset (\Delta^*_{MIX})(\pi(x^e))$, which proves (2.29). Similarly, we can prove (2.30). □

we are ready now to prove the main result of this section. For, we will assume that also the distributions $\mathcal{R}^e_i + \Delta^e_{MIX}$, $i = 1, \dots, \nu$, have constant dimension in a neighbourhood of x^e_0, with $\mathcal{R}^e_i$ and Δ^e_{MIX} as in proposition 2.9. Reasoning as in proposition 3.4.1, we prove that Δ^e_{MIX} and $\mathcal{R}^e_i + \Delta^e_{MIX}$ are involutive and invariant under $\widetilde{f}^e$, $\widetilde{g}^e_1, \dots, \widetilde{g}^e_{\nu+1}$. Thus, it makes sense to consider the restriction $\widetilde{f}^e|\mathcal{L}^{\Delta^e_{MIX}}_{x^e_0}$. Since $\pi_{*x^e}(\Delta^e_{MIX} + \mathcal{Q}^*)(x^e) = \Delta^*_{MIX}(\pi(x^e))$ (proposition 2.9), this restriction induces a dynamics, evolving on $\mathcal{L}^{\Delta^e_{MIX}}_{x^e_0}$, with trajectories being a subset of those of the system

$$\begin{aligned}
\dot{\zeta}_{\nu+3} &= \widetilde{f}_{\zeta_{\nu+3}}(0, \dots, 0, \zeta_{\nu+3}), \\
\dot{\zeta}_{\nu+4} &= \widetilde{f}_{\zeta_{\nu+4}}(0, \dots, 0, 0, \zeta_{\nu+3}, \zeta_{\nu+4}), \\
\dot{\zeta}_{\nu+5} &= \widetilde{f}_{\zeta_{\nu+5}}(0, \dots, 0, 0, \zeta_{\nu+3}, \zeta_{\nu+4}, \zeta_{\nu+5}), \\
\dot{w}^* &= \widetilde{f}_{w^*}(\zeta_{\nu+2}, \dots, \zeta_{\nu+5}, w^*),
\end{aligned} \tag{2.32}$$

where $(\zeta_{\nu+3}, \zeta_{\nu+4}, \zeta_{\nu+5}, w^*)$ is a set of local coordinates for $\mathcal{L}^{\Delta^e_{MIX}}_{x^e_0}$ and $(\zeta_{\nu+3}, \zeta_{\nu+4}, \zeta_{\nu+5})$ are chosen as in the proof of proposition 2.5. Since $\dot{x}^e = \widetilde{f}^e(x^e)$ is locally asymptotically stable in x^e_0 by assumption, it follows from (2.29) and by direct inspection of (2.32) that, in particular, the dynamics (2.3)*, given by $\dot{\zeta}_{\nu+3} = \widetilde{f}_{\zeta_{\nu+3}}(0, \dots, 0, 0, \zeta_{\nu+3})$, is locally asymptotically stable at the origin. Similarly, it makes sense to consider the restrictions $\widetilde{f}^e|\mathcal{L}^{\mathcal{R}^e_i+\Delta^e_{MIX}}_{x^e_0}$ and $\widetilde{g}^e_i|\mathcal{L}^{\mathcal{R}^e_i+\Delta^e_{MIX}}_{x^e_0}$ (the latter since $\widetilde{g}^e_i \in \mathcal{R}^e_i$). Since $\pi_{*x^e}(\mathcal{R}^e_i+\Delta^e_{MIX})(x^e) = (\Delta^*_{MIX}+\mathcal{R}^*_i)(\pi(x^e))$ (proposition 2.9), these restrictions induces a dynamics, evolving on $\mathcal{L}^{\mathcal{R}^e_i+\Delta^e_{MIX}}_{x^e_0}$, with trajectories being a subset of those of the system

$$\begin{aligned}
\dot{\xi}_2 &= \widetilde{f}_{\xi_2}(0, \xi_2) + \widetilde{g}^1_{\xi_2,i}(0, \xi_2)\alpha^1_i(0, \xi_2, \dots, \xi_5, \varphi^*), \\
\dot{\xi}_3 &= \widetilde{f}_{\xi_3}(0, \xi_3) + \sum_{h=1}^{2} \widetilde{g}^h_{\xi_3,i}(0, \xi_3)\alpha^h_i(0, \xi_2, \dots, \xi_5, \varphi^*), \\
\dot{\xi}_4 &= \widetilde{f}_{\xi_4}(0, \xi_2, \xi_3, \xi_4) + \sum_{h=1}^{2} \widetilde{g}^h_{\xi_4,i}(0, \xi_2, \xi_3, \xi_4)\alpha^h_i(0, \xi_2, \dots, \xi_5, \varphi^*), \\
\dot{\xi}_5 &= \widetilde{f}_{\xi_5}(0, \xi_2, \dots, \xi_5) + \sum_{h=1}^{2} \widetilde{g}^h_{\xi_5,i}(0, \xi_2, \dots, \xi_5)\alpha^h_i(0, \xi_2, \dots, \xi_5, \varphi^*), \\
\dot{\varphi}^* &= \widetilde{f}_{\varphi^*}(\xi_2, \dots, \xi_5, \varphi^*),
\end{aligned} \tag{2.33}$$

where $(\xi_2, \dots, \xi_5, \varphi^*)$ is a set of local coordinates for $\mathcal{L}^{\mathcal{R}^e_i+\Delta^e_{MIX}}_{x^e_0}$, $(\xi_2, \dots, \xi_5)$ are chosen as in the proof of proposition 2.5 and $\alpha^1_1(x, w), \alpha^2_1(x, w), \dots, \alpha^1_\nu(x, w), \alpha^2_\nu(x, w), \alpha_{\nu+1}(x, w)$

is the partition of $\alpha(x,w)$ corresponding to $u_1^1, u_1^2, \ldots, u_\nu^1, u_\nu^2, u_{\nu+1}$. Since $\dot{x}^e = \widetilde{f}^e(x^e)$ is locally asymptotically stable in x_0^e by assumption, it follows from (2.30) and by direct inspection of (2.33) that the dynamics (2.4)*, given by

$$\dot{\xi}_2 = \widetilde{f}_{\xi_2}(0,\xi_2) + \widetilde{g}^1_{\xi_2,i}(0,\xi_2)u_i^1,$$

is locally asymptotically stabilizable at the origin via dynamic feedback.

Conversely, suppose that the dynamics (2.3)* is locally asymptotically stable at the origin, the dynamics (2.4)* is locally asymptotically stabilizable at the origin via dynamic feedback and, in addition, the dynamics (1.11), associated to $\mathcal{R}_1^*, \ldots, \mathcal{R}_\nu^*$, (which will be denoted by (1.11)* in the sequel) is locally asymptotically stabilizable at the origin via dynamic feedback. Under these assumptions, NLSD is solvable. Indeed, reasoning as in section 2.1.3, we can assume without loss of generality that $\mathcal{Q}^* = 0$. From here, we can proceed as in section 3.4.2, once we prove that the distributions span $\{\widetilde{g}_i\}$, $i = 1, \ldots, \nu$, are linearly independent at each x of a neighbourhood of x_0. For, suppose that in each neighbourhood of x_0 there exists $\bar{x}$ such that

$$(\mathrm{span}\{\widetilde{g}_i\}(\bar{x})) \cap (\sum_{j\neq i} \mathrm{span}\{\widetilde{g}_j\}(\bar{x})) \neq 0$$

for some $i \in \{1, \ldots, \nu\}$. Since $\mathcal{Q}^* = 0$, $\mathcal{R}_1^*, \ldots, \mathcal{R}_\nu^*$ is the only regular solution. Thus, $\mathcal{R}_i^* \cap \mathcal{G} = \mathrm{span}\{\widetilde{g}_i\}$, which implies that

$$(\mathrm{span}\{\widetilde{g}_i\}(\bar{x})) \cap (\sum_{j\neq i} \mathrm{span}\{\widetilde{g}_j\}(\bar{x})) \subset (\mathcal{R}_i^* \cap \mathcal{G})(\bar{x}) \cap (\sum_{j\neq i} (\mathcal{R}_j^* \cap \mathcal{G}))(\bar{x}) \subset$$
$$\subset (\mathcal{R}_i^* \cap \sum_{j\neq i} \mathcal{R}_j^*)(\bar{x}) \cap \mathcal{G}(\bar{x}) \subset (\mathcal{R}^* \cap \mathcal{G})(\bar{x}) = (\mathcal{Q}^* \cap \mathcal{G})(\bar{x}).$$

This clearly gives a contradiction, since $\mathcal{Q}^* = 0$ for all $\bar{x}$ in a neighbourhood of x_0.

We sum up the above results in the following theorem.

Theorem 2.10. Assume that (A1), (A2) and (A3) hold. Moreover, assume that $\mathcal{R}_1^*, \ldots, \mathcal{R}_\nu^*$ is a regular set and consider the class of invertible dynamic feedback laws. If

a) the dynamics (2.3) is locally asymptotically stable at the origin,*

b) the dynamics (2.4) is locally asymptotically stabilizable at the origin via dynamic feedback,*

c) the dynamics (1.11) is locally asymptotically stabilizable at the origin via dynamic feedback,* *then NLSD is solvable. Conversely, assume that S^{e*} is regularly computable at x_{i0}^e. Moreover, the distributions $\mathcal{R}_i^e$, $i = 1, \ldots, \nu$, be regularly computable at x_0^e and the distributions $\mathcal{R}_i^e + \Delta_{MIX}^e$, $i = 1, \ldots, \nu$, and Δ_{MIX}^e have constant dimension in a neighbourhood of x_0^e. If NLSD is solvable, then a) and b) are satisfied.* □

From proposition 1.3 and theorem 2.10 we obtain the following important byproduct.

Corollary 2.11. Assume that (A1), (A2) and (A3) hold. Moreover, assume that $\mathcal{R}_1^*, \ldots, \mathcal{R}_\nu^*$ is a regular set and consider the class of invertible dynamic feedback. In addition, $\mathcal{Q}^* = 0$. If

a) the dynamics (2.3) is locally asymptotically stable at the origin,*

b) the dynamics (2.4) is locally asymptotically stabilizable at the origin via dynamic feedback,*

then NLSD is solvable. Conversely, assume that S^{e*} is regularly computable at x_{i0}^e. Moreover, the distributions $\mathcal{R}_i^e$, $i = 1, \ldots, \nu$, and Δ_{MIX}^e be regularly computable at x_0^e and the distributions $\mathcal{R}_i^e + \Delta_{MIX}^e$, $i = 1, \ldots, \nu$, and Δ_{MIX}^e have constant dimension in a neighbourhood of x_0^e. If NLSD is solvable, then a) and b) are satisfied. □

It is not hard to construct an example for which NLSD is solvable but the dynamics $(1.11)^*$ is *not* asymptotically stabilizable at the origin via *dynamic* feedback. Toward this end, we consider the following noninteractive system

$$\begin{aligned}
\dot{x}_i &= u_i, \qquad i = 1, 2, \\
\dot{x}_3 &= x_1 + x_2 + x_3, \\
\dot{x}_4 &= x_5^2 - x_2, \\
\dot{x}_5 &= u_3, \\
y_i &= x_i, \qquad i = 1, 2.
\end{aligned}$$

From straightforward computation, we obtain

$$\mathcal{R}_1^* = \text{span}\{\begin{pmatrix} 1 & 0 & 0 & 0 \\ 0 & 0 & 0 & 0 \\ 0 & 1 & 0 & 0 \\ 0 & 0 & 1 & 0 \\ 0 & 0 & 0 & 1 \end{pmatrix}\}, \ \mathcal{R}_2^* = \text{span}\{\begin{pmatrix} 0 & 0 & 0 & 0 \\ 1 & 0 & 0 & 0 \\ 0 & 1 & 0 & 0 \\ 0 & 0 & 1 & 0 \\ 0 & 0 & 0 & 1 \end{pmatrix}\}, \ \mathcal{Q}^* = \text{span}\{\begin{pmatrix} 0 & 0 \\ 0 & 0 \\ 0 & 0 \\ 1 & 0 \\ 0 & 1 \end{pmatrix}\}.$$

The dynamics $(1.11)^*$, associated to the maximal regular solution, is given by

$$\begin{aligned}
\dot{x}_4 &= x_5^2, \\
\dot{x}_5 &= u_3,
\end{aligned}$$

and it is *not* locally asymptotically stabilizable at the origin via dynamic feedback (even by means of continuous feedback: indeed, $\dot{x}_4$ is always positive). Nonetheless, NLSD is solvable. Indeed, note that the dynamics (1.11), associated to the regular solution

$$\mathcal{R}_1 = \text{span}\{\begin{pmatrix} 1 & 0 \\ 0 & 0 \\ 0 & 1 \\ 0 & 0 \\ 0 & 0 \end{pmatrix}\}, \ \mathcal{R}_2 = \mathcal{R}_2^*,$$

is locally asymptotically stabilizable at the origin via static feedback laws (since $\mathcal{Q} = 0$) and that the dynamics (2.3) and (2.4), associated to the regular solution $\mathcal{R}_1, \mathcal{R}_2$, are locally exponentially stable at the origin and, respectively, locally exponentially stabilizable at the origin via static feedback. Reasoning as in the proof of theorem 2.10 (with $\mathcal{R}_1^*, \mathcal{R}_2^*$ replaced by $\mathcal{R}_1$ and $\mathcal{R}_2$, respectively), we can easily prove our claim.

Remark 2.12. In the case that $\mathcal{R}_1^*, \ldots, \mathcal{R}_\nu^*$ is not a regular solution or, equivalently, (1.1) has not vector relative degree in x_0, we might conjecture that, as long as ND is solvable and in analogy with the nonlinear systems considered in chapter **3**, there exists a *canonical* dynamic extension of (1.1), which satisfies assumption (A1). If this is case, in order to solve NLSD, first we extend canonically (1.1) (when possible) and finally apply to the dynamically extended system the results of section 4.2. However, the possibility of doing this is still an open question. □

5.2.3 The case of block–partitioned outputs

The formulation of the problem of achieving noninteraction with stability, as given in section **3**.2, becomes meaningless if we consider nonlinear systems (0.1) with block–partitioned outputs. A reasonable and more general formulation is the following. Assume that there exists a regular feedback law (0.2) such that (0.1)–(0.2) is noninteractive. From proposition 1.1 it follows that there exist ν controllability distributions $\mathcal{R}_1, \ldots, \mathcal{R}_\nu$, a static feedback law (0.2) and a column (block) partition $v_1, \ldots, v_{\nu+1}$ of v which satisfy (1.3) and (1.4). Conversely, if there exist ν controllability distributions $\mathcal{R}_1, \ldots, \mathcal{R}_\nu$, a regular feedback law (0.2) and a (block) partition $v_1, \ldots, v_{\nu+1}$ of v which satisfy (1.3) and (1.4), then (0.1)–(0.2) is noninteractive.

Besides the noninteraction property, one may require also that the i–th output (block) of (0.1)–(0.2) be (locally) *controllable* through the i-th input (block) in the following sense (see [61]). Since $\mathcal{R}_i = \langle \tilde{f}, \tilde{g}_1, \ldots, \tilde{g}_{\nu+1} | \text{span}\{\tilde{g}_i, \tilde{g}_{\nu+1}\}\rangle$, if $\mathcal{R}_i$ has constant dimension, it is also involutive and $\mathcal{L}_{x_0}^{\mathcal{R}_i}$ is well-defined. It follows that the set $\mathcal{H}_i$ of the points $p \in \mathbb{R}^n$, which can be reached at time $t \geq 0$ from $x_0 = 0$ through the inputs of the i-th block, contains at least an open subset of $\mathcal{L}_{x_0}^{\mathcal{R}_i}$ ([83]). Thus, since $dh_i(0)$ has full rank, the image $h_i(\mathcal{H}_i)$ contains at least an open subset of $\mathbb{R}^{p_i}$. This property is referred to as (local) *output controllability*.

The property of output controllability can be immediately translated in geometric terms by saying that $\langle dh_i, \mathcal{R}_i\rangle(x) = \mathbb{R}^{p_i}$ ($\langle dh_i, \mathcal{R}_i\rangle(x)$ is the subspace of $\mathbb{R}^{p_i}$ spanned by the vectors $\langle dh_i, \tau\rangle(x)$ with $\tau \in \mathcal{R}_i$). Equivalently,

$$\mathcal{R}_i + \mathcal{K}_i = \mathbb{R}^n. \tag{3.1}$$

The above property is usually referred to as (local) *output controllability* property.

If we consider the class of static state–feedback laws, on the base of the above characterization, we can formulate the problem of achieving noninteraction and stability as follows.

Noninteracting control with stability via static state–feedback (NLSS). *Find, if possible, ν controllability distributions $\mathcal{R}_1, \ldots, \mathcal{R}_\nu$ for (0.1), a static feedback law (0.2) and a (block) partition $v_1, \ldots, v_{\nu+1}$ of v which satisfy (1.3), (1.4), (3.1) and such that the system (0.1)–(0.2) is locally asymptotically stable in $x = 0$.*

Noninteracting control with stability via dynamic state–feedback (NLSD). *Find, if possible, n^w, ν distributions $\mathcal{R}_1^e, \ldots, \mathcal{R}_\nu^e$, a feedback law (0.3) and a (block) partition $v_1, \ldots, v_{\nu+1}$ of v such that*

$$\mathcal{R}_i^e \subset \bigcap_{j \neq i} \mathcal{K}_j^e, \tag{3.2}$$

$$\mathcal{R}_i^e = \langle \tilde{f}^e, \tilde{g}_1^e, \ldots, \tilde{g}_{\nu+1}^e | span\{\tilde{g}_i^e, \tilde{g}_{\nu+1}^e\}\rangle, \tag{3.3}$$

$$\mathcal{R}_i^e + \mathcal{K}_i^e = \mathbb{R}^{n+n^w}, \tag{3.4}$$

and (0.1)–(0.3) is asymptotically stable in $(x^T, w^T)^T = (0^T, 0^T)^T$.

In general, the distributions $\mathcal{R}_1^e, \ldots, \mathcal{R}_\nu^e$ are *not* controllability distribution for the extended system.

In analogy with the previous sections, we will use the notations NS and ND to denote, respectively, the problem of noninteracting control (without internal stability) via static state–feedback and the problem of noninteracting control (without internal stability) via dynamic state–feedback.

If (A1) is replaced by the assumption that $\mathcal{R}_1^*, \ldots, \mathcal{R}_\nu^*$ is a regular solution, we can easily obtain a generalization of theorem 1.10 and corollary 1.11 to the case of block–partitioned outputs. On the other hand, we can easily prove also the following result, which generalizes theorem 2.10. Let $\widetilde{\Sigma}$ be the system obtained from (0.1) by applying a feedback law (0.2) which satisfies (1.4) with $\mathcal{R}_i = \mathcal{R}_i^*$ and for some (block) partition $v_1, \ldots, v_{\nu+1}$ of v.

Theorem 3.1. Assume that $\mathcal{R}_1^*, \ldots, \mathcal{R}_\nu^*$ is a regular solution. Moreover, assume that (A2) and (A3) hold and consider the class of weakly noninteractive feedback law. If

a) the dynamics (2.3) is locally asymptotically stable at the origin,*

b) the dynamics (2.4) is locally asymptotically stabilizable at the origin via dynamic feedback,*

c) the dynamics (1.11) is locally asymptotically stabilizable at the origin via dynamic feedback,*

then NLSD is solvable for $\widetilde{\Sigma}$. Conversely, assume that NLSD is solvable for $\widetilde{\Sigma}$. If, in addition, the distributions $\mathcal{R}_i^e = \langle \widetilde{f}^e, \widetilde{g}_1^e, \ldots, \widetilde{g}_{\nu+1}^e | \mathrm{span}\{\widetilde{g}_i^e, \widetilde{g}_{\nu+1}^e\}\rangle$, $i = 1, \ldots, \nu$, are regularly computable at x_0^e, $\mathcal{S}^{e*}$ is regularly computable at x_{i0}^e and the distributions $\mathcal{R}_i^e + \Delta_{MIX}^e$, $i = 1, \ldots, \nu$, and Δ_{MIX}^e have constant dimension in a neighbourhood of x_0^e, then, a) and b) are satisfied. □

Proof. The necessity follows as in theorem 2.10, since lemma 2.7 is in this case automatically satisfied. On the other hand, under assumptions a), b) and c), an invertible feedback law which solves NLSD can be constructed exactly as in theorem 3.4.17. It is easy to see that this feedback law is also weakly noninteractive. □

If $\mathcal{Q}^* = 0$, we obtain a generalized version of corollary 2.11. Although theorem 3.1 essentially restates the results of theorem 2.10 in a more general setting, the problem of achieving noninteraction and stability in the case of block–partitioned outputs is still far from being completely solved.

Before ending this section, we wish to remark that, in the linear case, every feedback law, which solves ND, can be expressed as the cascade of a *regular* feedback law together with a *weakly invertible* feedback law. This can be easily proved in the following way. It is sufficient to show that $\Delta^{e*} = \mathcal{V}_j^* + \mathcal{G}^w$ (see lemma 2.7). By definition, Δ^{e*} is the maximal subspace of $\mathbb{R}^{n+n^w}$ such that

$$[\widetilde{f}^e, \Delta^{e*}] \subset \Delta^{e*} + \mathcal{G}^e \tag{3.5}$$

and is contained in $\bigcap_{i \neq j} \mathcal{K}_i^e$. But (3.5) is equivalent to

$$[f^e, \Delta^{e*}] \subset \Delta^{e*} + \mathcal{G}^e. \tag{3.6}$$

It follows that Δ^{e*} is also the maximal subspace of $\mathbb{R}^{n+n^w}$ which satisfies (3.6) and is contained in $\bigcap_{i \neq j} \mathcal{K}_i^e$. This immediately implies that $\Delta^{e*} = \mathcal{V}_j^* + \mathcal{G}^w$.

Moreover, this proof suggests the following sufficient condition for Δ^{e*} being equal to $\mathcal{V}_j^* + \mathcal{G}^w$ for general nonlinear systems (0.1)

$$L_{\tilde{g}_i^e}(\Omega_{k-1}^e \cap (\mathcal{G}^e)^\perp) \subset \Omega_{k-1}^e$$

(see lemma 2.7).

CHAPTER 6

ISSUES IN NONINTERACTING CONTROL VIA OUTPUT–FEEDBACK

In the previous chapters, we have implicitly assumed the state available for feedback. Since this is not a common situation in practice, it might be desirable to assume available for feedback *only* a nonlinear function of the state (*measurement feedback*). For simplicity, we will assume throughout this chapter that this function is the output vector $h(x)$. Moreover, we will consider systems of the form (**3**.1.1) (further generalizations to more general systems can be obtained by using the results of chapter **5**) and feedback laws with the following structure

$$\begin{aligned} u &= \alpha(h(x), w) + \beta(h(x), w)v, \\ \dot{w} &= \gamma(h(x), w) + \delta(h(x), w)v, \end{aligned} \tag{1}$$

Correspondingly, the problem of achieving noninteraction and stability can be formulated as follows.

Noninteracting control problem with local stability via static output–feedback laws (NLSO). *Find, if possible, a static feedback law (1), defined in a neighbourhood of* x_0*, such that the system (**3**.1.1)–(1) is noninteractive, locally asymptotically stable in* x_0 *and has vector relative degree at* x_0.

Noninteracting control problem with local stability via dynamic output–feedback laws (NLSDO). *Find, if possible,* $n^w \geq 0$ *and a dynamic feedback law (1), defined in a neighbourhood of* (x_0, w_0)*, such that the system (**3**.1.1)–(1) is noninteractive, locally asymptotically stable in* (x_0, w_0) *and has vector relative degree at* (x_0, w_0).

In what follows, we will denote by NO and NDO the noninteracting control problem via static output–feedback (without the requirement of stability) and, respectively, the noninteracting control problem via dynamic output–feedback (without the requirement of stability).

In the *linear* case, as far as *regular* feedback is concerned, the problem of achieving noninteraction via *static* output–feedback is solvable if and only if

a) (**2**.1.1) has vector relative degree,

b) $A(\mathcal{R}_i^* \cap \ker\{dh\}) \subset \mathcal{R}_i^*, \qquad i = 1, \ldots, m.$

Indeed, a) is a necessary and sufficient condition for achieving noninteraction via regular *static state*-feedback (see assumption (AL1), remark **3**.3.6) and, as it will be shown in a moment, b) is a necessary and sufficient condition for each subspace $\mathcal{R}_i^*$ to be rendered invariant through regular *static output*-feedback. The necessity of b) is straightforward. On the other hand, b) implies that, if $u = F_i x$ is a feedback law which renders invariant $\mathcal{R}_i^*$, then there exists Γ_i such that $F_i = \Gamma_i C$. This amounts to the existence of a *static output*-feedback law which renders invariant $\mathcal{R}_i^*$.

If, besides noninteraction, one requires also internal stability, necessary and sufficient conditions can be easily derived, but we will not do this here at least for the case of static output feedback (see theorem **6**.1.2).

On the other hand, as far as *dynamic* output feedback is concerned, under the assumption that the system (**2**.1.1) is both controllable and observable (actually, stabilizability and detectability is enough!) and along the lines of the proof given by Morse & Wohnam in [89], we can easily prove that the problem of achieving noninteraction and stability via *dynamic output*-feedback is *equivalent* to the problem of achieving noninteraction via *dynamic state*-feedback. In other words, asymptotic stability can be always achieved, once noninteraction can be achieved. This is a little further step beyond what we already know about noninteracting control and can be easily seen in the following way.

Clearly, if noninteraction and stability can be achieved via *dynamic output*-feedback then noninteraction can be also achieved via *dynamic state*-feedback. To prove the converse, we will construct a dynamic output–feedback law, which achieves noninteraction and stability or, equivalently, we will find m controllability subspaces $\mathcal{R}_1^e, \ldots, \mathcal{R}_\nu^e$ for (**2**.1.7), a feedback law $u^e = F^e x^e + G^e v^e$ and a column (block) partition $G_1^e, \ldots, G_{\nu+1}^e$ of G^e, which satisfy (**2**.1.8)–(**2**.1.10) and such that (**2**.1.7), with $u^e = F^e x^e + G^e v^e$, is asymptotically stable. The construction is more or less the same as the one proposed in remark **3**.4.14 for nonlinear systems. We will assume that (**2**.1.1) is asymptotically stable (if not, since (**2**.1.1) is both observable and controllable, we can always asymptotically stabilize it through *dynamic output*-feedback).

Let $\dim \mathcal{R}_i^* \cap \mathcal{B} = q_i$, $\dim \mathcal{R}_i^* = s_i$ and F_i and G_i be $m \times n$ and $m \times m$ matrices, with $G_i = (\, G_{i1} \quad G_{i2} \,)$ invertible and G_{i1} a $m \times q_i$ matrix, such that $\mathcal{R}_i^* = \langle A + BF_i | \mathrm{Im}\{BG_{i1}\} \rangle$, $\mathcal{R}_i^* \cap \mathcal{B} = \mathrm{Im}\{BG_{i1}\}$ and $(A + BF_i)|\mathcal{R}_i^*$ is Hurwitz. Moreover, let $n^w = \sum_{j=1}^{m} s_m$ and let x^e,

A^e, B^e and C^e be as in section **2**.1. After possibly changing coordinates, we have

$$A = \begin{pmatrix} A_{i1} & A_{i2} \\ A_{i3} & A_{i4} \end{pmatrix}, \; BG_i = \begin{pmatrix} B_{i1} & B_{i2} \\ 0_{(n-s_i)\times q_i} & B_{i4} \end{pmatrix}, \; G_i^{-1}F_i = \begin{pmatrix} F_{i1} & F_{i2} \\ F_{i3} & F_{i4} \end{pmatrix},$$

with A_{i1} a $s_i \times s_i$ matrix and B_{i1} a $s_i \times q_i$ matrix, $A_{i3}+B_{i4}F_{i3} = 0$ and $A_{i1}+B_{i1}F_{i1}+B_{i2}F_{i3}$ is Hurwitz. Note that in general the subspaces span $\{BG_1\}, \ldots,$ span $\{BG_m\}$ are not linearly independent (this is the case, for example, when (**2**.1.1) has not vector relative degree). Correspondingly, define

$$G^e = \begin{pmatrix} G_{11} & G_{21} & \cdots & G_{m1} \\ B_{11} & 0_{s_1\times q_2} & \cdots & 0_{s_1\times q_m} \\ \vdots & \vdots & \cdots & \vdots \\ 0_{s_m\times q_1} & 0_{s_m\times q_2} & \cdots & B_{m1} \end{pmatrix}$$

and

$$F^e = \begin{pmatrix} 0_{m\times n} & G_{11}F_{11} + G_{12}F_{13} & \cdots & G_{m1}F_{m1} + G_{m2}F_{m3} \\ 0_{s_1\times n} & A_{11} + B_{11}F_{11} + B_{12}F_{13} & \cdots & 0_{s_1\times s_m} \\ \vdots & \vdots & \vdots & \vdots \\ 0_{s_m\times n} & 0_{s_m\times s_1} & \cdots & A_{m1} + B_{m1}F_{m1} + B_{m2}F_{m3} \end{pmatrix}.$$

It is straightforward calculation to check that the subspaces $\mathcal{R}_i^e = \langle A^e + B^e F^e | \text{span}\{B^e G_i^e\}\rangle$, $i = 1, \ldots, m$, with G_i^e the matrix having as columns the columns $\sum_{j=1}^{i-1} q_j + 1, \ldots, \sum_{j=1}^{i} q_j$ of G^e, respectively, satisfy (**2**.1.8)–(**2**.1.10). Moreover, since (**2**.1.1) is asymptotically stable and each matrix $(A + BF_i)|\mathcal{R}_i^* = A_{i1} + B_{i1}F_{i1} + B_{i2}F_{i3}$ is Hurwitz, it follows that the system (**2**.1.7), with $u^e = F^e x^e + G^e v^e$, is also asymptotically stable. Since the first n columns of F^e are identically zero, the feedback law (2) is an *output* feedback law we look for.

In the case of *nonlinear* systems, we could be tempted to believe that, in analogy with b), the condition

$$\begin{aligned} &[f, \mathcal{R}_i^* \cap \ker\{dh\}] \subset \mathcal{R}_i^*, \\ &[g_j, \mathcal{R}_i^* \cap \ker\{dh\}] \subset \mathcal{R}_i^*, \qquad j = 1, \ldots, m, \end{aligned} \tag{2}$$

for all x in a neighbourhood of x_0 (see [52]), is enough for the existence of a static feedback law (1) which renders invariant the distribution $\mathcal{R}_i^*$. Counterexamples show that this is not true in general. Indeed, an interesting obstruction is pointed out by the following example

$$\begin{aligned} \dot{x}_{11} &= x_{12} \\ \dot{x}_{12} &= a_0(x_{12})x_{22} + u_1 \\ \dot{x}_{22} &= u_2 \\ \dot{x}_3 &= x_{11} + x_{22} - x_3 \\ y_i &= x_{ii}, \qquad i = 1, 2, \end{aligned} \tag{3}$$

with $a_0(x_{12})$ a smooth *nonconstant* function of the *unmeasured* state–variable x_{12}. Indeed, we have

$$\mathcal{R}_1^* = \text{span}\{\frac{\partial}{\partial x_{11}}, \frac{\partial}{\partial x_{12}}, \frac{\partial}{\partial x_3}\},$$
$$\mathcal{R}_2^* = \text{span}\{\frac{\partial}{\partial x_{22}}, \frac{\partial}{\partial x_3}\}.$$

Note that (3) has vector relative degree and that (2) is clearly satisfied. However, it is not possible to find a static output–feedback law which achieves noninteraction, since for *no* smooth $\alpha_1(x_{11}, x_{22})$ the function $\frac{\partial}{\partial x_{22}}(a_0(x_{12})x_{22} + \alpha_1(x_{11}, x_{22}))$ is identically zero in a neighbourhood of x_0. On the other hand, if $a_0(x_{12}) = 1$, the system becomes *linear* and NLSO is always solvable. A characterization in geometric terms of the obstruction, pointed out by example (3), is the object of the next section.

6.1 Noninteracting control via static output–feedback

In this section we will investigate necessary and sufficient conditions for solving NO via *invertible* feedback laws (see section **3.3** for remarks on the class of feedback laws considered). We will consider the class of nonlinear systems (**3**.1.1) satisfying assumption (A1), (A2) and (A3) (generalizations are possible for wider classes of systems). Assumption (A1) is quite natural in this context, since it is a necessary condition for solving NO.

Let us apply the *state*–feedback law (**3**.3.1) to (**3**.1.1). Let $\mathcal{L}$ be the Lie ideal generated by the vector field $\tilde{g}_1, \ldots, \tilde{g}_m$ in the Lie algebra generated by the vector fields $\tilde{f}, \tilde{g}_1, \ldots, \tilde{g}_m$. Moreover, if $\{r_1, \ldots, r_m\}$ is the vector relative degree of (**3**.1.1), let $\bar{\mathcal{L}}$ be the set consisting of all the vector fields of $\mathcal{L}$ but the vector fields $ad_{\tilde{f}}^{r_1-1}\tilde{g}_1, \ldots, ad_{\tilde{f}}^{r_m-1}\tilde{g}_m$. The vector fields of $\mathcal{L}$ have a particular structure. Indeed, span $\{\bar{\mathcal{L}}\} = \ker\{dh\}$ and, by definition of vector relative degree, each function $L_{ad_{\tilde{f}}^{r_i-1}\tilde{g}_i} h_i(x)$ is nonzero at each x in a neighbourhood of x_0. This particular structure will be useful in the subsequent analysis.

For a given smooth distribution Δ_1 and vector field τ, once a smooth distribution Δ_2 is chosen in such a way that $\Delta_1 \oplus \Delta_2 = \mathbb{R}^n$, we will denote by τ_{Δ_1} (τ_{Δ_2}) the vector field which assigns to each x in a neighbourhood of x_0 the projection of $\tau(x)$ along $\Delta_2(x)$ ($\Delta_1(x)$). Choose $\mathcal{Q}_i$ such that $(\sum_{j \neq i} \mathcal{R}_j^*) \oplus \mathcal{Q}_i = \mathbb{R}^n$. Moreover, in what follows, for a given feedback law $u = \alpha(h(x)) + \beta(h(x))v$, we let $\tilde{f}(x) = f(x) + G(x)\alpha(x)$ and $\tilde{g}_j(x) = G(x)\beta_j(h(x))$ with $\beta_j(x)$ the j–th column of $\beta(x)$.

Proposition 1.1. Assume that (A1), (A2) and (A3) hold. Assume in addition that the distributions $\langle \tilde{f}, \tilde{g}_1, \ldots, \tilde{g}_m | \text{span}\{\tilde{g}_i\}\rangle$, $i = 1, \ldots, m$, *have constant dimension in a neigh-*

bourhood of x_0. *Then, if NO is solvable, for each* $i \in \{1,\ldots,m\}$ *we have*

$$[[f,Z],X] \subset \mathrm{span}\{[g_l,Z] : l=1,\ldots,m\} + \sum_{j\neq i} \mathcal{R}_j^*, \tag{1.1}$$

$$[[g_s Z],X] \subset \mathrm{span}\{[g_l,Z] : l=1,\ldots,m\} + \sum_{j\neq i} \mathcal{R}_j^*, \qquad s=1,\ldots,m, \tag{1.2}$$

for each $Z \in \bar{\mathcal{L}}$, *for all* $X \in \sum_{j\neq i} \mathcal{R}_j^*$ *and for all* x *in a neighbourhood of* x_0,

$$[f, \mathcal{R}_i^* \cap \ker\{dh\}] \subset \mathcal{R}_i^*,$$
$$[g_s, \mathcal{R}_i^* \cap \ker\{dh\}] \subset \mathcal{R}_i^*, \qquad s=1,\ldots,m,$$

for all x *in a neighbourhood of* x_0, *and there exists an* $m \times m$ *matrix* $\beta(h(x))$, *invertible at each* x *in a neighbourhood of* x_0, *such that for all* x *in a neighbourhood of* x_0

$$G\beta_i \in \mathcal{R}_i^*, \tag{1.3}$$

where β_i *denotes the i–th column of* β.

Conversely, assume that $(\sum_{j\neq i} \mathcal{R}_j^* + \mathcal{G}) \cap \mathrm{span}\{[g_l,Z] : l=1,\ldots,m\} = 0$ *for each* $Z \in \bar{\mathcal{L}}$ *and that* $\mathrm{span}\{[g_l,Z] : l=1,\ldots,m\}$ *has constant dimension in a neighbourhood of* x_0. *If (2) and (1.1)-(1.2) hold and there exists an* $m \times m$ *matrix* $\beta(h(x))$, *invertible at each* x *in a neighbourhood of* x_0, *such that for all* x *in a neighbourhood of* x_0 *(1.3) holds, then NO is solvable.* □

Proof. Assume that there exists a feedback law $u = \alpha(h(x)) + \beta(h(x))v$ which solves NO. Conditions (2) and (1.3) are an easy consequence of the invertibility of $\beta(h(x))$ at each x of a neighbourhood of x_0.

Now, we will prove (1.1) and (1.2). To this end, we will assume that $g_i \in \mathcal{R}_i^*$. As a consequence, $\beta(h(x))$ is a *diagonal* matrix with entries $\beta_{11}(h(x)),\ldots,\beta_{mm}(h(x))$ nonzero at each x in a neighbourhood of x_0. Let $\mathcal{G}_i = \mathrm{span}\{g_i\}$ for $i=1,\ldots,m$ (for simplicity, we can choose $\mathcal{Q}_i$ in such a way that $\mathcal{Q}_i \supset \mathcal{G}_i$, but the forthcoming results hold independently of this choice). Since $(\sum_{j\in I} \mathcal{R}_j^*) \cap \mathcal{G} = \mathrm{span}\{g_j : j \in I\}$ with $I \subset \{1,\ldots,m\}$ (see [64]), we obtain

$$\sum_{j\neq i} \mathcal{R}_j^* + \mathcal{G} = \sum_{j\neq i} \mathcal{R}_j^* \oplus \mathcal{G}_i,$$

where $\oplus$ denotes at each x the direct sum of subspaces. Let $X_{1i},\ldots,X_{l_i i}$ be vector fields, linearly independent at each x in a neighbourhood of x_0 and such that $\sum_{j\neq i} \mathcal{R}_j^* = \mathrm{span}\{X_{1i},\ldots,X_{l_i i}\}$. Since by proposition **3.3.1** $[g_l\beta_{ll}, X_{hi}] \in \sum_{j\neq i} \mathcal{R}_j^*$ for all l and

$$[g_i\beta_{ii}, X_{hi}](x) = \beta_{ii}(h(x))[g_i, X_{hi}](x) - (L_{X_{hi}}\beta_{ii}(h(x)))g_i(x),$$

we have necessarily $(\beta_{ii}(h(x))c_{hi}(x))g_i(x) = (L_{X_{hi}}\beta_{ii}(h(x)))g_i(x)$, where the c_{hi}'s are unique smooth real-valued functions such that $[g_i, X_{hi}](x) = c_{hi}(x)g_i(x) + \sum_{k=1}^{l_i} d_{ki}(x)X_{ki}(x)$. Equivalently,

$$\beta_{ii}(h(x))c_{hi}(x) = L_{X_{hi}}\beta_{ii}((h(x))). \tag{1.4}$$

Moreover,

$$L_Z\beta_{ii}((h(x))) = 0, \qquad Z \in \bar{\mathcal{L}}. \tag{1.5}$$

Note that the distributions $\sum_{j\neq i}\mathcal{R}_j^* + \ker\{dh\}$. $i = 1,\ldots,m$, are involutive, since ker $\{dh\} = \bar{\mathcal{L}}$ and $\sum_{j\neq i}\mathcal{R}_j^*$ is by construction invariant under $\tilde{f}, \tilde{g}_1, \ldots, \tilde{g}_m$ and, thus, under any vector field of $\bar{\mathcal{L}}$.

A necessary and sufficient condition to solve the set of partial differential equations (1.4)–(1.5) is (see [53])

$$L_Z c_{hi}(x) = \sum_{k=1}^{l_i} c_{ki}(x)b_{ki}(x), \tag{1.6}$$

where the b_{ki}'s are unique smooth real-valued functions such that $[X_{hi}, Z](x) = \sum_{k=1}^{l_i} b_{ki}(x)$ $X_{ki}(x)$. Note that for all $Z \in \bar{\mathcal{L}}$

$$-[[g_i, X_{hi}], Z](x) + [[g_i, Z], X_{hi}](x) + [g_i, [X_{hi}, Z]](x) = 0 \tag{1.7}$$

and that by construction span $\{\bar{\mathcal{L}}\} = \ker\{dh\}$. Since $[X_{hi}, Z] \in \sum_{j\neq i}\mathcal{R}_j^*$, from (1.7) we obtain

$$[[g_i, X_{hi}]_{\mathcal{G}_i}, Z]_{\mathcal{Q}_i}(x) + [[g_i, Z]_{\mathcal{Q}_i}, X_{hi}]_{\mathcal{Q}_i}(x) + [g_i, [X_{hi}, Z]]_{\mathcal{Q}_i}(x) = 0 \tag{1.8}$$

and, from (1.6) together with (1.8),

$$-c_{hi}(x)[g_i, Z]_{\mathcal{Q}_i}(x) + [[g_i, Z]_{\mathcal{Q}_i}, X_{hi}]_{\mathcal{Q}_i}(x) = 0. \tag{1.9}$$

From (2), (1.9) and since $[g_j, Z]_{\mathcal{Q}_i}(x) = 0$ for $j \neq i$, (1.2) follows immediately. From $[f, X_{hi}](x) = c_{0hi}(x)g_i(x) + \sum_{k=1}^{l_i} d_{0ki}(x)X_{ki}(x)$, also (1.1) follows in a similar way.

Finally, we will prove that (2) and (1.1)–(1.3) are also sufficient to solve NO. For, by assumption there exists a $m \times m$ matrix $\beta(h(x))$, invertible at each x in a neighbourhood of x_0, such that, if β_i is the i–th column of β, we have $G\beta_i \in \mathcal{R}_i^*$. For simplicity of notation, further on we will assume directly that $g_i \in \mathcal{R}_i^*$. From (1.8), since $(\sum_{j\neq i}\mathcal{R}_j^* + \mathcal{G}) \cap$ span$\{[g_l, Z] : l = 1,\ldots,m\} = 0$ and $[[g_j, Z], X_{hi}] \in \sum_{j\neq i}\mathcal{R}_j^*$, we obtain (1.6). This implies the existence of a smooth function $\alpha_i(h(x))$ such that $\alpha_i(h(x_0)) = 0$ and $[f+g_i\alpha_i, \sum_{j\neq i}\mathcal{R}_j^*] \subset \sum_{j\neq i}\mathcal{R}_j^*$. Following the same lines of [64], it is easy to prove that the feedback law

$$\alpha(h(x)) = \begin{pmatrix} \alpha_1(h(x)) \\ \vdots \\ \alpha_m(h(x)) \end{pmatrix}, \quad \beta(h(x)) = I_{m\times m}$$

solves NO. □

The necessary conditions (1.1)–(1.2) are trivially satisfied for linear systems, so that (2) is the only nontrivial obstruction, as already known. Note that, as expected, (1.2) is not satisfied for (3), since $[g_j, Z](x) = 0$ for all j and there exists a vector field $\tau \in \bar{\mathcal{L}}$ having the form $\begin{pmatrix} 0 \\ 1 \\ 0 \\ 0 \end{pmatrix}$ so that $[[f,\tau], X_2] \notin \mathcal{R}_2^*$ with $X_2 = \begin{pmatrix} 0 \\ 0 \\ 1 \\ 0 \end{pmatrix}$.

Along the lines of section **3**.3, we can easily derive some necessary and sufficient conditions to achieve *both* noninteraction and stability. To this end, assume that NLSO is solvable. Let us apply to (**3**.1.1) any *output* feedback law which solves NLSO and denote by $\tilde{\Sigma}$ the system correspondingly obtained. Under assumptions (A1), (A2) and (A3), following the proof of proposition **3**.3.3, one can find a change of coordinates $z(x) = (\, z_1^T(x) \;\; \cdots \;\; z_{m+1}^T(x)\,)^T$, $z(x_0) = 0$, defined in a neighbourhood of x_0 and such that $\tilde{\Sigma}$ is expressed in the form (**3**.3.8) (for simplicity of notation, in (**3**.3.8) we will let $z = x$ and $v = u$). In analogy with section **3**.3, it is possible to define the nonlinear dynamics

$$\dot{x}_{m+1} = \tilde{f}_{m+1}(0,\ldots,0,x_{m+1}) \tag{1.10}$$

and

$$\dot{x}_i = \tilde{f}(x_i) + \tilde{g}_{ii}(x_i)v_i, \qquad i = 1,\ldots,m. \tag{1.11}$$

The dynamics (1.10) and (1.11) are *uniquely* (in some sense to be specified) associated to a given system (**3**.1.1), which satisfies assumptions (A1), (A2) and (A3). This renders the definition of (1.10) and (1.11) *independent* (in some sense) of the particular output–feedback law chosen to obtain (**3**.3.8). Given the system (**3**.3.8), let us apply to it any *output* feedback law $u = \alpha'(h(x)) + \beta'(h(x))v$ which solves NLSO. Denote by $\tilde{\Sigma}'$ the system obtained from (**3**.3.8) and by (1.10)′ and (1.11)′ the systems defined on $\tilde{\Sigma}'$ and corresponding to (1.10) and (1.11), respectively. In analogy with proposition **3**.3.4, it is possible to show that the dynamics (1.10) and (1.10)′ are *equal* up to changes of coordinates, defined in a neighbourhood of x_0 and, moreover, the systems (1.11) and (1.11)′ are *equal* up to changes of coordinates and invertible *output*–feedback transformations, defined in a neighbourhood of x_0.

Following section **3**.3, one can state the following theorem.

Theorem 1.2. Assume that (A1), (A2) and (A3) hold and consider the class of invertible feedback laws. If NO is solvable and, in addition,

a) the dynamics (1.10) is locally asymptotically stable at the origin,

b) each system (1.11) is locally asymptotically stabilizable at the origin via static output–feedback,

then NLSO is solvable. Conversely, assume, in addition, that the distributions $\langle \widetilde{f}, \widetilde{g}_1, \dots, \widetilde{g}_m | \operatorname{span}\{\widetilde{g}_i\}\rangle$, $i = 1, \dots, m$, *have constant dimension in a neighbourhood of* x_0. *If NLSO is solvable, then a) and b) are satisfied.* □

We want to remark that, in the case of linear systems, contrary to the case of static state–feedback, condition b) is not automatically satisfied in general.

6.2 Noninteracting control via dynamic output–feedback: examples and counterexamples

Assume for a moment that either (2) or (1.1) or (1.2) do not hold for a system satisfying assumption (A1). We know from proposition 1.1 that NO is not solvable. On the other hand, we might ask if *dynamic* output–feedback can help overcoming the obstructions (2), (1.1) and (1.2). Through some worked examples, we will show that dynamic output–feedback is indeed helpful in overcoming the obstruction (2), but, in general, not (1.1)–(1.2). However, at the moment, necessary and sufficient conditions for solving even NDO are still an open question and are a challenging object for further research.

We will give two examples: the first one shows that (2) is *not* necessary to solve NDO and gives a possible procedure for constructing solutions of NLSDO for systems having vector relative degree at x_0, the other one proves that possible obstructions, for which NLSDO might not be solvable (even if NLSD is), may be caused simply by the fact that either (1.1) or (1.2) is not satisfied.

Let us begin with the following example

$$\begin{aligned}
\dot{x}_1 &= x_3 b_1(y_1) + u_1 \\
\dot{x}_2 &= x_3 b_2(y_2) + u_2 \\
\dot{x}_3 &= a_1(x_1) + a_2(x_2) + x_3 \\
\dot{x}_4 &= c(x_1, x_2, x_3, x_4) \\
y_i &= x_i, \qquad i = 1, 2,
\end{aligned} \tag{2.1}$$

with $a_i(x_i)$, $b_i(y_i)$ and $c(x_1, \ldots, x_4)$ smooth functions of their arguments, $a_i(0) = 0$, $\frac{\partial a_i}{\partial x_i}(0) \neq 0$, $i = 1, 2$, $c(0,0,0,0) = 0$, $\frac{\partial^{(2)} c}{\partial x_1 \partial x_2}(0,0,0,0) \neq 0$. Moreover, assume that $\dot{x}_4 = c(0,0,0,x_4)$ is locally asymptotically stable at $x_4 = 0$.

Note that (2.1) is locally *exponentially* stabilizable at $x = 0$ through *static state-*feedback. Note also that (2.1) has vector relative degree $\{1,1\}$. Since after a state-feedback $u_1 = -x_3 b_1(y_1) + v_1$, $u_2 = -x_3 b_2(y_2) + v_2$, the resulting system is noninteractive and satisfies the necessary and sufficient conditions of theorem **3**.4.17, NLSD is solvable (see example (**3**.5.1) for the construction of the feedback law which solves NLSD). Note that NLSS is not solvable, as a consequence of theorem **3**.3.5 and NO is not solvable since (2) is not satisfied. Note also that (1.1)–(1.2) are trivially satisfied. One might ask if NLSDO is solvable for (2.1). This is answered in the next proposition.

Proposition 2.1. The problem of achieving noninteraction with local stability via dynamic output feedback is solvable for (2.1). □

Proof. By straightforward computations we obtain in a neighbourhood of x_0

$$\begin{aligned}
\Delta_{MIX} &= \mathrm{span}\{\partial/\partial x_4\}, \\
\mathcal{R}^* &= \mathrm{span}\{\partial/\partial x_3, \partial/\partial x_4\}, \\
\mathcal{R}_i^* &= \mathrm{span}\{\partial/\partial x_i, \partial/\partial x_3, \partial/\partial x_4\} \qquad i = 1, 2.
\end{aligned}$$

By assumption, the dynamics

$$\dot{x}_4 = c(0,0,0,x_4) \tag{2.2}$$

is locally asymptotically stable at $x_4 = 0$ and each system

$$\begin{aligned}
\dot{x}_i &= u_i \\
\dot{x}_3 &= a_i(x_i) + x_3
\end{aligned} \tag{2.3}$$

is locally *exponentially* stabilizable at the origin via *static state*–feedback law. Let $n^w = 2$ and define

$$X_{11}^e(x^e) = \begin{pmatrix} 1 \\ 0 \\ 0 \\ 0 \\ 0 \end{pmatrix}, X_{12}^e(x^e) = \begin{pmatrix} 0 \\ 0 \\ 1 \\ 1 \\ 0 \end{pmatrix}, X_{21}^e(x^e) = \begin{pmatrix} 0 \\ 1 \\ 0 \\ 0 \\ 0 \end{pmatrix}, X_{22}^e(x^e) = \begin{pmatrix} 0 \\ 0 \\ 1 \\ 0 \\ 1 \end{pmatrix},$$

with $\mathcal{R}_i^e = \text{span}\,\{X_{ij}^e : j = 1,2\}$. Note that $\mathcal{R}_i^e \subset \bigcap_{j \neq i} \ker\{dh_j\} \oplus \mathcal{G}^w$, where $\oplus$ denotes at each point (x, w) the external direct sum of subspaces. We have constructed the vector fields X_{ik}^e according to the procedure of section 3.4.3. It can be easily checked that the output feedback law

$$\begin{aligned} &\alpha^e(x^e) = (\,-(\mu_1+\mu_2)b_1(y_1) \quad -(\mu_1+\mu_2)b_2(y_2) \quad (a_1(x_1)+\mu_1) \quad (a_2(x_2)+\mu_2)\,)^T, \\ &\beta^e(x^e) = I_{4\times 4} \end{aligned} \tag{2.4}$$

renders locally invariant the distributions $\mathcal{R}_1^e$ and $\mathcal{R}_2^e$ and, in addition, denoting by β_i^e the i–th column of β^e, is such that $G\beta_i^e \in \mathcal{R}_i^e$ for $i = 1,2$. After a linear change of coordinates, the closed–loop system (2.1)–(2.4) can be transformed by the global change of coordinates

$$\begin{aligned} z_1^e &= x_1,\, z_2^e = \mu_1, \\ z_3^e &= x_2,\, z_4^e = \mu_2, \\ z_5^e &= x_3 - \mu_1 - \mu_2, \end{aligned}$$

into a system of the form

$$\begin{aligned} \dot z_1^e &= u_1 + z_5^e b_1(z_1^e) \\ \dot z_2^e &= a_1(z_1^e) + \mu_1 + u_1^w \\ \dot z_3^e &= u_2 + z_5^e b_2(z_3^e) \\ \dot z_4^e &= a_2(z_3^e) + \mu_2 + u_2^w \\ \dot z_5^e &= z_5^e - u_1^w - u_2^w \\ y_1 &= z_1^e \\ y_2 &= z_3^e. \end{aligned}$$

Note that, under our assumptions, the system $\dot z_5^e = z_5^e - u_1^w - u_2^w$ can be locally asymptotically stabilized through a feedback law of the form

$$\begin{aligned} \dot\xi_1 &= F_1(\xi_1, \xi_2), \\ \dot\xi_2 &= F_2(\xi_1, \xi_2), \\ u_1^w &= -a_1(\xi_1), \\ u_2^w &= -a_2(\xi_2). \end{aligned}$$

We conclude that a feedback law which solves NLSDO is given by

$$u_i = -(\mu_1 + \mu_2) b_i(y_i) + H_{i1} y_i + H_{i3}\mu_i + v_i, \qquad i = 1, 2,$$
$$\dot{\xi}_i = F_i(\xi_1, \xi_2), \qquad i = 1, 2,$$
$$\dot{\mu}_i = a_i(y_i) + \mu_i - a_i(\xi_i), \qquad i = 1, 2,$$

where H_{i1} and H_{i3} are real numbers such that the matrix

$$\begin{pmatrix} H_{i1} & H_{i3} \\ \frac{\partial a_i}{\partial x_i}(0) & 1 \end{pmatrix}$$

is Hurwitz (the existence of H_{i1} and H_{i3} is guaranteed by the controllability of the linear approximation of (2.1) in a neighbourhood of the origin). □

It is important to note the two following facts. First, under our assumptions the dynamics (2.2), which is exactly the dynamics of (2.1) on $\mathcal{L}_{x_0}^{\Delta_{MIX}}$, is locally asymptotically stable at $x_4 = 0$. We know that, under some regularity assumptions, this condition is necessary to solve NLSD (and, *a fortiori*, NLSDO) for (2.1).

Another necessary condition for solving NLSDO is that each system (2.3) be locally asymptotically stabilizable at the origin through *dynamic* feedback (not necessarily output feedback): this is a consequence of theorem **3**.3.17. From the proof of proposition 1 we see that this condition is also sufficient. □

Next, let us consider the system

$$\begin{aligned} \dot{x}_1 &= x_3^2 + u_1 \\ \dot{x}_2 &= u_2 \\ \dot{x}_3 &= x_1 + x_2 + x_3 \\ y_i &= x_i, \qquad i = 1, 2, \end{aligned} \tag{2.5}$$

and let us investigate the solvability of NDO for (2.5). First of all, note that, after applying a feedback law $u_1 = -x_3^2 + v_1$, the resulting system satisfies the necessary and sufficient conditions of theorem **3**.4.17 and, thus, NLSD is solvable. Note that NLSS is not solvable, as a consequence of theorem **3**.3.17. Moreover, note that (1.1) is not satisfied. We want to prove that this is an obstruction for solving even NDO. As a matter of fact, we obtain the following result.

Proposition 2.2. The problem of achieving noninteraction via dynamic output feedback is not solvable for (2.5). □

Proof. In a neighbourhood of $x = 0$ we have

$$\mathcal{R}^* = \text{span}\{\partial/\partial x_3\},$$
$$\mathcal{R}_i^* = \text{span}\{\partial/\partial x_i, \partial/\partial x_3\}, \qquad i = 1, 2.$$

Assume that there exists a dynamic output feedback law (1) which solves NDO. We will denote, as usual, $\begin{pmatrix} f + G\alpha \\ \gamma \end{pmatrix}$ by $\widetilde{f}^e$ and $\begin{pmatrix} G\beta_j \\ \delta_j \end{pmatrix}$ by $\widetilde{g}_j^e$ for $j = 1, \ldots, m$. Let $\mathcal{L}_i$ be the Lie ideal generated by the vector field g_i in the Lie algebra generated by the vector fields f, $g_1, \ldots, g_m$ and $\mathcal{L}_i^e$ be the Lie ideal generated by the vector field $\widetilde{g}_i^e$ in the Lie algebra generated by the vector fields $\widetilde{f}^e$, $\widetilde{g}_1^e, \ldots, \widetilde{g}_m^e$. Moreover, let $y = \begin{pmatrix} y_1 \\ y_2 \end{pmatrix}$ and π be the natural projection $\pi(x^e) = x$, with π_{*x^e} its differential at x^e. By arguments similar to those of proposition **3**.4.4, one can easily prove the following fact.

Fact # 2.1. $\beta_{ij}(y, w) = 0$ for $j \neq i$ and for all (y, w) in a neighbourhood of $(0, 0)$; moreover,

$$\pi_{*x^e}(\text{span}\{\mathcal{L}_i^e\})(x^e) = (\text{span}\{\frac{\partial}{\partial x_i}, \frac{\partial}{\partial x_3}\})(\pi(x^e)) \tag{2.6}$$

for all x^e in a neighbourhood of x_0^e.

As a consequence, we have

$$\widetilde{f}^e(x^e) = \begin{pmatrix} x_3^2 + \alpha_1(y, w) \\ \alpha_2(y, w) \\ x_1 + x_2 + x_3 \\ \delta(y, w) \end{pmatrix}, \quad \widetilde{g}_1^e(x^e) = \begin{pmatrix} \beta_{11}(y, w) \\ 0 \\ 0 \\ \gamma_1(y, w) \end{pmatrix}, \quad \widetilde{g}_2^e(x^e) = \begin{pmatrix} 0 \\ \beta_{22}(y, w) \\ 0 \\ \gamma_2(y, w) \end{pmatrix}.$$

The closed–loop system (2.5)–(1) has, by assumption, vector relative degree $\{r_1^e, r_2^e\}$ at x_0^e. Moreover, (2.5) itself has vector relative degree $\{1, 1\}$ at x_0. The following property is an easy consequence of the vector relative degree property.

Fact # 2.2. Either one of following facts may happen:

a) $r_2^e = 1$ and $\beta_{22}(y, w) \neq 0$ at each (y, w) in a neighbourhood of $(0, 0)$;

b) $r_2^e > 1$ and $\beta_{22}(y, w) = 0$ for all (y, w) in an open neighbourhood of $(0, 0)$, $L_{\widetilde{g}_2^e} L_{\widetilde{f}^e}^k \alpha_2(x^e) = 0$ for $0 \leq k < r_2^e - 2$ and for all x^e in an open neighbourhood of x_0^e; moreover, $L_{\widetilde{g}_2^e} L_{\widetilde{f}^e}^{r_2^e - 2} \alpha_2(x^e) \neq 0$ at each x^e in a neighbourhood of x_0^e.

Indeed, if $r_2^e = 1$, since the closed–loop system (2.5)–(1) is noninteractive and has vector relative degree $\{r_1^e, r_2^e\}$ at x_0^e and $L_{g_2} h_2(x) = 1$, then necessarily $\beta_{22}(y, w) \neq 0$ for all (y, w)

in a neighbourhood of (0,0), i.e. a). On the other hand, if $r_2^e > 1$ and since $L_{g_2}h_2(x) = 1$ and $L_X h_2(x) = 0$ for any $X \in (\mathcal{L}_1 + \mathcal{L}_2)/\{g_2\}$, we have

$$L_{ad^k_{\tilde{f}^e}\tilde{g}^e_2} h_2(x^e) = 0, \qquad 0 \le k < r_2^e - 1$$

for all x^e in a neighbourhood of x_0^e and (up to signs)

$$L_{ad^{r_2^e-1}_{\tilde{f}^e}\tilde{g}^e_2} h_2(x^e) = (L_{g_2}h_2(x)) L_{ad^{r_2^e-2}_{\tilde{f}^e}\tilde{g}^e_2}\alpha_2(x^e) = L_{ad^{r_2^e-2}_{\tilde{f}^e}\tilde{g}^e_2}\alpha_2(x^e)$$

for all x^e in a neighbourhood of x_0^e. This immediately implies b).

By induction, it is possible to show that

$$ad^k_{\tilde{f}^e}\tilde{g}^e_2(x^e) = \begin{pmatrix} 0 \\ 0 \\ 0 \\ \varphi_k(x^e) \end{pmatrix}, \qquad 0 \le k < r_2^e - 1, \tag{2.7}$$

where

$$\varphi_k(x^e) = \sum_{j=0}^{2k} a_j(y,w) x_3^j$$

for some smooth (possibly vector) functions $a_j(y,w)$ such that $a_1(0,0) = 0$. Indeed, the first component of $ad^k_{\tilde{f}^e}\tilde{g}^e_2(x^e)$, $0 \le k < r_2^e - 1$, is identically zero by fact #2.1. Moreover, its second component is equal to $\beta_{22}(y,w)$ if $k = 0$ or to $L_{ad^{k-1}_{\tilde{f}^e}\tilde{g}^e_2}\alpha_2(x^e)$ if $k \ge 1$. Thus, the second component is identically zero by fact #2.2. Finally, by direct computations, the third component is easily seen to be identically zero, since the first two are. For $k = 0$ $\varphi_1(x^e)$ has the claimed form by fact #2.2. Now, assume that $\varphi_k(x^e)$ has the above form for some $k < r_2^e - 2$. Since $\alpha_1(0,0) = \alpha_2(0,0) = 0$, $\delta(0,0) = 0$ and by induction hypothesis,

$$\varphi_{k+1}(x^e) = \{\sum_{j=0}^{2k} \frac{\partial a_j}{\partial y_1}(y,w)x_3^j\}(x_3^2 + \alpha_1(y,w)) + \{\sum_{j=0}^{2k} \frac{\partial a_j}{\partial y_2}(y,w)x_3^j\}\alpha_2(y,w) +$$

$$+ \{\sum_{j=0}^{2k} \frac{\partial a_j}{\partial w}(y,w)x_3^j\}\delta(y,w) + \sum_{j=1}^{2k} j a_j(y,w)x_3^{j-1}(x_1 + x_2 + x_3) - \frac{\partial \delta}{\partial w}(y,w)\varphi_k(x^e) =$$

$$= \sum_{j=0}^{2(k+1)} \bar{a}_j(y,w)x_3^j$$

for some smooth (possibly vector) functions $\bar{a}_j(y,w)$ such that $\bar{a}_1(0,0) = 0$. Moreover, from straightforward computations and fact #2.2, we obtain

$$ad^{r_2^e-1}_{\tilde{f}^e}\tilde{g}^e_2(x^e) = \begin{pmatrix} 0 \\ \psi_{r_2^e-1}(x^e) \\ 0 \\ \varphi_{r_2^e-1}(x^e) \end{pmatrix},$$

where $\varphi_{r_2^e-1}(x^e) = \sum_{j=0}^{2(r_2^e-1)} \lambda_j(y,w)x_3^j$ and $\psi_{r_2^e-1}(x^e) = \sum_{j=0}^{2(r_2^e-1)} \mu_j(y,w)x_3^j$ for some smooth (possibly vector) functions $\lambda_j(y,w)$ and $\mu_j(y,w)$ such that $\lambda_1(0,0)$ and $\mu_1(0,0)=0$. Since $\varphi_{r_2^e-1}(x^e)$ is equal (up to signs) to $\beta_{22}(y,w)$ if $r_2^e=1$ or to $L_{ad_{\tilde{f}^e}^{r_2^e-2}\tilde{g}_2^e}\alpha_2(x^e)$ if $r_2^e>1$, by fact #2.2 we have $\varphi_{r_2^e-1}(x^e)\neq 0$ for all x^e in a neighbourhood of x_0^e.

Moreover,

$$ad_{\tilde{f}^e}^{r_2^e}\tilde{g}_2^e(x^e) = \begin{pmatrix} 0 \\ \psi_{r_2^e}(x^e) \\ -\psi_{r_2^e-1}(x^e) \\ \varphi_{r_2^e}(x^e) \end{pmatrix},$$

where $\varphi_{r_2^e}(x^e) = \sum_{j=0}^{2r_2^e} c_j(y,w)x_3^j$ and $\psi_{r_2^e}(x^e) = \sum_{j=0}^{2r_2^e} d_j(y,w)x_3^j$ for some smooth (possibly vector) functions $c_j(y,w)$ and $d_j(y,w)$ such that $c_1(0,0)=d_1(0,0)=0$. From fact # 2.1, it follows that $\pi_*(ad_{\tilde{f}^e}^k\tilde{g}_2^e)(x^e)$, $k\geq 0$, is a vector of $\mathcal{R}_2^*(\pi(x^e))$. In particular, the first component of $ad_{\tilde{f}^e}^{r_2^e+1}\tilde{g}_2^e$ must be identically zero or, equivalently,

$$0 = L_{ad_{\tilde{f}^e}^{r_2^e}\tilde{g}_2^e}(x_3^2+\alpha_1(y,w)) = \frac{\partial\alpha_1}{\partial y_2}(y,w)\psi_{r_2^e}(x^e) - 2x_3\psi_{r_2^e-1}(x^e) + \frac{\partial\alpha_1}{\partial w}(y,w)\varphi_{r_2^e}(x^e) \quad (2.8)$$

for all x^e in a neighbourhood of x_0^e. In particular, (2.8) must hold for $(y,w)=(0,0)$, which clearly gives a contradiction.

□

BIBLIOGRAPHY

[1] G. Basile, G. Marro, Controlled and conditioned invariant subspaces in linear systems theory, *Journal of Optimization Theory and Applications*, vol. 3, 1969, 306–315.

[2] S. Battilotti, A sufficient condition for noninteracting control with stability via dynamic state-feedback, *IEEE Transactions on Automatic Control*, vol. 36, 1991, 1033–1045.

[3] S. Battilotti, A sufficient condition for noninteracting control with stability via dynamic state-feedback: block-partitioned outputs, *International Journal of Control*, vol. 55, 1992, 1141–1160.

[4] S. Battilotti, W.P. Dayawansa, Stabilization of globally noninteractive nonlinear systems via dynamic state-feedback, *Journal of Mathematical Systems, Estimation and Control*, vol.1, 1991, 441–463.

[5] S. Battilotti, W.P. Dayawansa, Necessary and sufficient conditions for noninteracting control with stability for a class of nonlinear systems, *Systems & Control Letters*, vol. 17, 1991, 327–338.

[6] S. Battilotti, Necessary conditions for nonlinear block noninteracting control with stability via dynamic state–feedback, *Systems & Control Letters*, vol. 19, 1992, 481–491.

[7] W.B. Boothby, *An introduction to differentiable manifolds and riemaniann geometry*, Academic Press, New York, 1975.

[8] W.M. Boothby, R. Marino, Dynamic feedback stabilization of a scalar system, preprint.

[9] R. Bott, Lectures on characteristic classes of foliations, *Lecture notes in Math*, 279, Springer Verlag, New York, Berlin, 1972, 1–80.

[10] R. Brockett, M.D. Mesarovic, The reproducibility of multivariable systems, *Journal of Mathematical Analysis and Application*, vol. 11, 1965, 548–563.

[11] C. Byrnes, A. Isidori, Global feedback stabilization of nonlinear systems, *Proceedings of the 24-th Conference on Decision and Control*, Lauderdale, 1985, 1031–1037.

[12] C. Byrnes, A. Isidori, Asymptotic stabilization of minimum phase systems, *IEEE Transactions on Automatic Control*, vol. 36, 1991, 1122–1137.

[13] D. Cheng, Design for noninteracting decomposition of nonlinear systems, *IEEE Transaction on Automatic Control* vol. 33, 1988, 1070–1074.

[14] W.I. Chow, Uber system von linearen partiellen differential gleichungen ester ordnung, *Mathematical Annales*, vol. 117, 1939, 98–105.

[15] D. Claude, Decoupling of nonlinear systems, *Systems & Control Letters*, vol. 1, 1982, 301–308.

[16] J.M. Coron, L. Praly, Adding an integrator for the stabilization problem, *Systems & Control Letters*, vol. 17, 1991, 89–104.

[17] M. Cremer, A precompensation of minimal order for decoupling a linear multivariable system, *International Journal of Control*, vol. 14, 1971, 1089–1103.

[18] P.S.P. Da Silva, On the decoupling by dynamic feedback of nonlinear affine smooth systems, *Proceedings of European Control Conference*, 1992, Bordeaux, 552–557.

[19] W.P. Dayawansa, W.B. Boothby, T.J. Tarn, D. Elliot, Global state and feedback equivalence of nonlinear systems, *Systems and Control Letters*, vol. 6, 1985, 229–234.

[20] W.P. Dayawansa, D. Cheng, W.B. Boothby, T.J. Tarn, On the global (f, g)–invariance of a class of nonlinear systems, *Proceedings of the 25-th Conference on Decision and Control*, Athens, 1986, 2065–2068.

[21] W.P. Dayawansa, D. Cheng, W.B. Boothby, T.J. Tarn, Global (f, g)–invariance of nonlinear systems, *SIAM Journal on Control and Optimization*, vol. 26, 1988, 1119–1131.

[22] W.P. Dayawansa, C.F. Martin, Asymptotic stabilization of two dimension real analytic systems, *Systems & Control Letters*, vol. 12, 1989, 205–211.

[23] W.P. Dayawansa, C.F. Martin, G. Knowles, Asymptotic stabilization of a class smooth two dimensional systems,*SIAM Journal on Control and Optimization*,1989.

[24] J. Descusse, J.M. Dion, On the structure at infinity of linear square decoupled systems, *IEEE Transaction on Automatic Control*, vol. 27, 1982, 971–974.

[25] J. Descusse, J.F. Lafay, M. Malabre, Further results on Morgan's problem, *Systems & Control Letters*, vol. 4, 1984, 203–208.

[26] J. Descusse,C.H. Moog, Decoupling with dynamic compensation for strong invertible affine systems, *International Journal of Control*, vol. 42, 1986, 1387–1398.

[27] J. Descusse,C.H. Moog, Dynamic decoupling for right invertible nonlinear systems, *Systems & Control Letters*, vol. 8, 1987, 345–349.

[28] J. Descusse, J.F. Lafay, M. Malabre, Solution to Morgan's problem, *IEEE Transactions on Automatic Control*, vol. 33, 1988, 732–739.

[29] J. Descusse, The block essential orders and the dynamic block decoupling problem, *Systems & Control Letters*, vol. 14, 1990, 153–159.

[30] M.D. Di Benedetto, J.W. Grizzle, C.H. Moog, Rank invariants for nonlinear systems, *SIAM Journal on Optimization and Control*, vol. 72, 1989, 658–672.

[31] P.L. Falb, W.A. Wolovich, Decoupling in the design and synthesis of multivariable control systems, *IEEE Transactions on Automatic Control*, vol. 12, 1967, 651–659.

[32] M. Fliess, A new approach to the noninteracting control problem in nonlinear system theory, *Proceedings of Allerton Conference on Communications, Control and Computation*, Monticello, 1985, 123–129.

[33] M. Fliess, Some remarks on nonlinear invertibility and dynamic state–feedback, *Proceedings of M.N.T.S.* (in *Theory and Applications of Nonlinear Control Systems*, C.I. Byrnes and A. Lindquist eds.), Stockholm, 1985, 115–123.

[34] M. Fliess, A new approach to the structure at infinity of nonlinear systems, *Systems & Control Letters*, vol. 7, 1986, 419–421.

[35] E. Freund, The structure of decoupled nonlinear systems, *International Journal of Control*, vol. 21, 1975, 443–450.

[36] E.G. Gilbert, The decoupling of multivariable systems by state–feedback, *SIAM Journal of Control and Optimization*, vol. 7, 1969, 50–63.

[37] A. Glumineau, C.H. Moog, Essential orders and the nonlinear decoupling problem, *International Journal of Control*, vol. 50, 1989, 1825–1834.

[38] L.C.J.M. Gras, H. Nijmeijer, Decoupling in nonlinear systems: from linearity to nonlinearity, *IEEE Proceedings*, vol. 136, 1989, 53–62.

[39] J.W. Grizzle, M.D. Di Benedetto, C.H. Moog, Computing the differential output rank of a nonlinear system, *Proceedings of the 26–th Conference on Decision and Control*, Los Angeles, 1987, 142–145.

[40] J.W. Grizzle , A. Isidori, Block noninteracting control with stability via static state–feedback, *Mathematics of Control, Systems and Signals*, vol. 2, 1989, 315–341.

[41] I.J. Ha, The standard decomposed system and noninteracting feedabck control of nonlinear systems, *SIAM Journal on Control and Optimization*, 1986.

[42] J. Ha, G. Gilbert, A complete characterization of decoupling control laws for general class of nonlinear systems, *IEEE Transactions on Automatic Control*, vol. 31, 1986, 823–830.

[43] W. Hahn, *Stability of motion*, Springer Verlag, New York, 1967.

[44] M.L.J. Hautus, M. Heymann, Linear Feedback Decoupling – Transfer Function Analysis, *IEEE Transactions on Automatic Control*, vol. 28, no. 8, 1983, 823–832.

[45] A.N. Herrera, J.F. Lafay, P. Zagalak, The extended noninteractor in the study of noninvertible control problems, *Proceedings of the 30–th Conference on Decision and Control*, Brighton, 1991.

[46] M. Heymann, Comments on "Pole assignment in multiinput controllable linear systems", *IEEE Transactions on Automatic Control*, vol. 13, 1968, 748–749.

[47] A.N. Herrera, J.F. Lafay, New results about the Morgan's problem, to appear in *IEEE Transaction on Automatic Control*.

[48] J. Hirshorn, Invertibility of multivariable control systems, *IEEE Transaction on Automatic Control*, vol. 24, 1979, 855–865.

[49] J. Huang, W.J. Rugh, Approximate noninteracting control with stability for nonlinear systems, *IEEE Transaction on Automatic Control*, 1991.

[50] H.J.C. Huijberts, A.J. van der Shaft, Input–output decoupling with stability for Hamiltonian systems, *Mathematical Control Signals and Systems*, vol. 2, 1989, 315–341.

[51] A. Isidori, Control of Nonlinear Systems via dynamic state–feedback, in *Algebraic and Geometric Methods in Nonlinear Control Theory*, M. Fliess and M.Hazewinkel eds., 1986, 121–145.

[52] A. Isidori, A.J. Krener, C. Gori Giorgi,S. Monaco, Nonlinear decoupling via feedback: a differential geometric approach, *IEEE Transactions on Automatic Control*, vol. 26, 1981, 331–345.

[53] A. Isidori, A.J. Krener, C. Gori Giorgi, S. Monaco, Locally (f, g) invariant distributions, *Systems & Control Letters*, vol. 1, 1981, 12–15.

[54] A. Isidori and J.W. Grizzle, Fixed modes and nonlinear noninteracting control with stability, *IEEE Trans. on Autom. Contr.*, 8, 1988, 345–349.

[55] A. Isidori, *Nonlinear Control Systems*, Springer Verlag, 2nd ed., 1989.

[56] S. Kobayashi, K. Nomizu, *Foundations of differential geometry*, John Wiley and Sons, 1963.

[57] C.H. Moog, Nonlinear Decoupling and Structure at Infinity, *Mathematics of Control, Signals and Systems*, vol. 1, 1988, 257–268.

[58] S. Morse, W.M. Wohnam, Status of noninteracting control, *IEEE Transactions on Automatic Control*, vol. 16, 1971, 568–581.

[59] H. Nijmeijer, Invertibility of affine nonlinear control systems: a geometric approach, *Systems & Control Letters*, vol. 2, 1982, 163–169.

[60] H. Nijmeijer, Feedback decomposition of nonlinear control systems, *IEEE Transactions on Automatic Control*, vol. 28, 1983, 861–862.

[61] H. Nijmeijer, The triangular decoupling problem for nonlinear control systems, *Nonlinear Analysis and Applications*, vol. 8, 1984, 273–279.

[62] H. Nijmeijer, J.M. Schumacher, Zeroes at infinity for affine nonlinear control systems, *IEEE Transactions on Automatic Control*, vol. 30, 1985, 566–573.

[63] H. Nijmeijer, J.M. Schumacher, On the input–output decoupling of nonlinear systems, *Proceedings of the Conference on Algebraic and Geometric Methods in Nonlinear Con-*

trol Theory, Paris, 1985.

[64] H. Nijmeijer, J.M. Schumacher, The regular local noninteracting control problem, *SIAM Journal on Control and Optimization*, vol. 8, 1986, 1232–1245.

[65] H. Nijmeijer, Right invertibility for a class of nonlinear control systems: a geometric approach, *Systems & Control Letters*, vol. 7, 1986, 125–132.

[66] H. Nijmeijer, W. Respondek, Decoupling via dynamic compensation for nonlinear control systems, *Proceedings of the 25–th Conference on Decision and Control*, Athens, 1986, 192–197.

[67] H. Nijmeijer, W. Respondek, Dynamic input–output decoupling of nonlinear control systems, *IEEE Transactions on Automatic Control*, vol. 33, 1988, 1065–1070.

[68] H. Nijmeijer, A. Van der Shaft, Controlled invariance of nonlinear systems, *IEEE Transactions on Automatic Control*, vol. 27, 1982, 904–914.

[69] W.M. Porter, Diagonalization and inverse for nonlinear systems, *International Journal of Control*, vol. 10, 1970, 67–76.

[70] W. Respondek, On decomposition of nonlinear control systems, *Systems & Control Letters*, vol. 1, 1982, 301–308.

[71] W. Respondek, Right and left invertibility of nonlinear control systems, *Proceedings of the Conference of Rutgers University* (in *Controllability and Optimal Control*, H.J. Sussmann ed.), M. Dekker, 1987.

[72] W. Respondek, H. Nijmeijer, On local right–invertibility of nonlinear control systems, *Control Theory and Advanced Technology*, vol. 4, 1988, 325–348.

[73] L.M. Silverman, H.J. Payne, Input–output structure of nonlinear systems with application to the decoupling problem, *SIAM Journal on Control and Optimization*, vol. 9, 1971, 199–233.

[74] L.M. Silverman, Inversion of linear systems, *IEEE Transactions on Automatic Control*, vol. 14, 1969, 270–276.

[75] S.N. Singh, W.J. Rugh, Decoupling in a class of nonlinear systems by state variable feedback, *ASME Journal on Dynamical Systems, Measurement and and Control*, vol. 94, 1972, 323–329.

[76] S.N. Singh, Decoupling of invertible nonlinear systems with state–feedback and precompensation, *IEEE Transactions on Automatic Control*, vol. 26, 1981, 331–345.

[77] S.N. Singh, A modified algorithm for invertibility in nonlinear systems, *IEEE Transactions on Automatic Control*, vol. 26, 1981, 595–598.

[78] P.K. Sinha, State–feedback decoupling of nonlinear systems, *IEEE Transaction on Automatic Control*, vol. 22, 1977, 487–489.

[79] E.D. Sontag, Remarks on stabilization and input–to–state stability, *Proceedings of the 28-th Conference on Decision and Control*, Tampa, 1989, 1975–1980.

[80] E.D. Sontag, Feedback stabilization of nonlinear systems, *Conference on Mathematical Theory of Networks and Systems*, 1989.

[81] E.D. Sontag, Further facts about input to state stability, *IEEE Transactions on Automatic Control*, vol. 35, 1990, 473–476.

[82] H.J. Sussmann, P. Kokotovic, The peaking phenomenon and the global stabilization of nonlinear systems, *IEEE Transactions on Automatic Control*, vol. 36, 1991, 424–440.

[83] H.J. Sussmann, V. Jurdjevic, Controllability of nonlinear systems, *Journal of Differential Equations*, vol. 12, 1972, 95–116.

[84] J. Tsinias, On the existence of control Lyapunov functions: generalizations of Vidyasagar's theorem on nonlinear stabilization, *SIAM Journal on Control and Oprimization*, vol. 30, 1992, 879–893.

[85] J. Tsinias, Existence of control Lyapunov functions and applications to state–feedback stabilizability of nonlinear systems, *SIAM Journal on Control and Oprimization*, vol. 29, 1991, 457–473.

[86] K. Wagner, Nonlinear noninteraction with stability by dynamic state–feedback, *SIAM Journal on Optimization and Control*, vol. 29, 1991, 609–622.

[87] K. Wagner, On nonlinear noninteraction with stability, *Proceedings of the 28-th Conference on Decision and Control*, Tampa, 1989, 1994–1999.

[88] S.H. Wang, Design of precompensator for decoupling problem, *Electronics Letters*, vol. 6, 1970, 739–741.

[89] W.M. Wohnam, S. Morse, Decoupling and pole assignment by dynamic compensation, *SIAM Journal on Control and Optimization*, vol. 8, 1970, 331–337.

[90] W.M. Wohnam, S. Morse, Decoupling and pole assignment in linear multivariable systems: a geometric approach, *SIAM Journal on Control and Optimization*, vol. 8, 1970, 1–18.

[91] W.M. Wohnam, *Linear Multivariable Control: A Geometric Approach*, 2nd ed., Springer Verlag, New York, 1979.

[92] W. Zhan, T.J. Tarn, A. Isidori, A canonical dynamic extension for noninteraction with stability for affine nonlinear square systems, *Systems & Control Letters*, vol. 17, 1991, 177–184.

INDEX

Lecture Notes in Control and Information Sciences

Edited by M. Thoma

1989–1993 Published Titles:

Vol. 135: Nijmeijer, Hendrik; Schumacher, Johannes M. (Eds.)
Three Decades of Mathematical System Theory. A Collection of Surveys at the Occasion of the 50th Birthday of Jan C. Willems.
562 pp. 1989 [3-540-51605-0]

Vol. 136: Zabczyk, Jerzy W. (Ed.)
Stochastic Systems and Optimization. Proceedings of the 6th IFIP WG 7.1 Working Conference, Warsaw, Poland, September 12-16, 1988.
374 pp. 1989 [3-540-51619-0]

Vol. 137: Shah, Sirish L.; Dumont, Guy (Eds.)
Adaptive Control Strategies for Industrial Use. Proceedings of a Workshop held in Kananaskis, Canada, 1988.
360 pp. 1989 [3-540-51869-X]

Vol. 138: McFarlane, Duncan C.; Glover, Keith
Robust Controller Design Using Normalized Coprime Factor Plant Descriptions.
206 pp. 1990 [3-540-51851-7]

Vol. 139: Hayward, Vincent; Khatib, Oussama (Eds.)
Experimental Robotics I. The First International Symposium, Montreal, June 19-21, 1989.
613 pp. 1990 [3-540-52182-8]

Vol. 140: Gajic, Zoran; Petkovski, Djordjija; Shen, Xuemin (Eds.)
Singularly Perturbed and Weakly Coupled Linear Control Systems. A Recursive Approach.
202 pp. 1990 [3-540-52333-2]

Vol. 141: Gutman, Shaul
Root Clustering in Parameter Space.
153 pp. 1990 [3-540-52361-8]

Vol. 142: Gündes, A. Nazli; Desoer, Charles A.
Algebraic Theory of Linear Feedback Systems with Full and Decentralized Compensators.
176 pp. 1990 [3-540-52476-2]

Vol. 143: Sebastian, H.-J.; Tammer, K. (Eds.)
System Modelling and Optimizaton. Proceedings of the 14th IFIP Conference, Leipzig, GDR, July 3-7, 1989.
960 pp. 1990 [3-540-52659-5]

Vol. 144: Bensoussan, A.; Lions, J.L. (Eds.)
Analysis and Optimization of Systems. Proceedings of the 9th International Conference. Antibes, June 12-15, 1990.
992 pp. 1990 [3-540-52630-7]

Vol. 145: Subrahmanyam, M. Bala
Optimal Control with a Worst-Case Performance Criterion and Applications.
133 pp. 1990 [3-540-52822-9]

Vol. 146: Mustafa, Denis; Glover, Keith
Minimum Enthropy H Control.
144 pp. 1990 [3-540-52947-0]

Vol. 147: Zolesio, J.P. (Ed.)
Stabilization of Flexible Structures. Third Working Conference, Montpellier, France, January 1989.
327 pp. 1991 [3-540-53161-0]

Vol. 148: Not published

Vol. 149: Hoffmann, Karl H; Krabs, Werner (Eds.)
Optimal Control of Partial Differential Equations. Proceedings of IFIP WG 7.2 - International Conference. Irsee, April, 9-12, 1990.
245 pp. 1991 [3-540-53591-8]

Vol. 150: Habets, Luc C.
Robust Stabilization in the Gap-topology.
126 pp. 1991 [3-540-53466-0]

Vol. 151: Skowronski, J.M.; Flashner, H.; Guttalu, R.S. (Eds.)
Mechanics and Control. Proceedings of the 3rd Workshop on Control Mechanics, in Honor of the 65th Birthday of George Leitmann, January 22-24, 1990, University of Southern California.
497 pp. 1991 [3-540-53517-9]

Vol. 152: Aplevich, J. Dwight
Implicit Linear Systems.
176 pp. 1991 [3-540-53537-3]

Vol. 153: Hajek, Otomar
Control Theory in the Plane.
269 pp. 1991 [3-540-53553-5]

Vol. 154: Kurzhanski, Alexander; Laseicka, Irena (Eds.)
Modelling and Inverse Problems of Control for Distributed Parameter Systems. Proceedings of IFIP WG 7.2 - IIASA Conference, Laxenburg, Austria, July 1989.
170 pp. 1991 [3-540-53583-7]

Vol. 155: Bouvet, Michel; Bienvenu, Georges (Eds.)
High-Resolution Methods in Underwater Acoustics.
244 pp. 1991 [3-540-53716-3]

Vol. 156: Hämäläinen, Raimo P.; Ehtamo, Harri K. (Eds.)
Differential Games - Developments in Modelling and Computation. Proceedings of the Fourth International Symposium on Differential Games and Applications, August 9-10, 1990, Helsinki University of Technology, Finland.
292 pp. 1991 [3-540-53787-2]

Vol. 157: Hämäläinen, Raimo P.; Ehtamo, Harri K. (Eds.)
Dynamic Games in Economic Analysis. Proceedings of the Fourth International Symposium on Differential Games and Applications. August 9-10, 1990, Helsinki University of Technology, Finland.
311 pp. 1991 [3-540-53785-6]

Vol. 158: Warwick, Kevin; Karny, Miroslav; Halouskova, Alena (Eds.)
Advanced Methods in Adaptive Control for Industrial Applications.
331 pp. 1991 [3-540-53835-6]

Vol. 159: Li, Xunjing; Yong, Jiongmin (Eds.)
Control Theory of Distributed Parameter Systems and Applications. Proceedings of the IFIP WG 7.2 Working Conference, Shanghai, China, May 6-9, 1990.
219 pp. 1991 [3-540-53894-1]

Vol. 160: Kokotovic, Petar V. (Ed.)
Foundations of Adaptive Control.
525 pp. 1991 [3-540-54020-2]

Vol. 161: Gerencser, L.; Caines, P.E. (Eds.)
Topics in Stochastic Systems: Modelling, Estimation and Adaptive Control.
1991 [3-540-54133-0]

Vol. 162: Canudas de Wit, C. (Ed.)
Advanced Robot Control. Proceedings of the International Workshop on Nonlinear and Adaptive Control: Issues in Robotics, Grenoble, France, November 21-23, 1990.
Approx. 330 pp. 1991 [3-540-54169-1]

Vol. 163: Mehrmann, Volker L.
The Autonomous Linear Quadratic Control Problem. Theory and Numerical Solution.
177 pp. 1991 [3-540-54170-5]

Vol. 164: Lasiecka, Irena; Triggiani, Roberto
Differential and Algebraic Riccati Equations with Application to Boundary/Point Control Problems: Continuous Theory and Approximation Theory.
160 pp. 1991 [3-540-54339-2]

Vol. 165: Jacob, Gerard; Lamnabhi-Lagarrigue, F. (Eds.)
Algebraic Computing in Control. Proceedings of the First European Conference, Paris, March 13-15, 1991.
384 pp. 1991 [3-540-54408-9]

Vol. 166: Wegen, Leonardus L. van der
Local Disturbance Decoupling with Stability for Nonlinear Systems.
135 pp. 1991 [3-540-54543-3]

Vol. 167: Rao, Ming
Integrated System for Intelligent Control.
133 pp. 1992 [3-540-54913-7]

Vol. 168: Dorato, Peter; Fortuna, Luigi; Muscato, Giovanni
Robust Control for Unstructured Perturbations: An Introduction.
118 pp. 1992 [3-540-54920-X]

Vol. 169: Kuntzevich, Vsevolod M.; Lychak, Michael
Guaranteed Estimates, Adaptation and Robustness in Control Systems.
209 pp. 1992 [3-540-54925-0]

Vol. 170: Skowronski, Janislaw M.; Flashner, Henryk; Guttalu, Ramesh S. (Eds.)
Mechanics and Control. Proceedings of the 4th Workshop on Control Mechanics, January 21-23, 1991, University of Southern California, USA.
302 pp. 1992 [3-540-54954-4]

Vol. 171: Stefanidis, P.; Paplinski, A.P.; Gibbard, M.J.
Numerical Operations with Polynomial Matrices: Application to Multi-Variable Dynamic Compensator Design.
206 pp. 1992 [3-540-54992-7]

Vol. 172: Tolle, H.; Ersü, E.
Neurocontrol: Learning Control Systems Inspired by Neuronal Architectures and Human Problem Solving Strategies.
220 pp. 1992 [3-540-55057-7]

Vol. 173: Krabs, W.
On Moment Theory and Controllability of Non-Dimensional Vibrating Systems and Heating Processes.
174 pp. 1992 [3-540-55102-6]

Vol. 174: Beulens, A.J. (Ed.)
Optimization-Based Computer-Aided Modelling and Design. Proceedings of the First Working Conference of the New IFIP TC 7.6 Working Group, The Hague, The Netherlands, 1991.
268 pp. 1992 [3-540-55135-2]

Vol. 175: Rogers, E.T.A.; Owens, D.H.
Stability Analysis for Linear Repetitive Processes.
197 pp. 1992 [3-540-55264-2]

Vol. 176: Rozovskii, B.L.; Sowers, R.B. (Eds.)
Stochastic Partial Differential Equations and their Applications. Proceedings of IFIP WG 7.1 International Conference, June 6-8, 1991, University of North Carolina at Charlotte, USA.
251 pp. 1992 [3-540-55292-8]

Vol. 177: Karatzas, I.; Ocone, D. (Eds.)
Applied Stochastic Analysis. Proceedings of a US-French Workshop, Rutgers University, New Brunswick, N.J., April 29-May 2, 1991.
317 pp. 1992 [3-540-55296-0]

Vol. 178: Zolésio, J.P. (Ed.)
Boundary Control and Boundary Variation. Proceedings of IFIP WG 7.2 Conference, Sophia- Antipolis,France, October 15-17, 1990.
392 pp. 1992 [3-540-55351-7]

Vol. 179: Jiang, Z.H.; Schaufelberger, W.
Block Pulse Functions and Their Applications in Control Systems.
237 pp. 1992 [3-540-55369-X]

Vol. 180: Kall, P. (Ed.)
System Modelling and Optimization. Proceedings of the 15th IFIP Conference, Zurich, Switzerland, September 2-6, 1991.
969 pp. 1992 [3-540-55577-3]

Vol. 181: Drane, C.R.
Positioning Systems - A Unified Approach.
168 pp. 1992 [3-540-55850-0]

Vol. 182: Hagenauer, J. (Ed.)
Advanced Methods for Satellite and Deep Space Communications. Proceedings of an International Seminar Organized by Deutsche Forschungsanstalt für Luft-und Raumfahrt (DLR), Bonn, Germany, September 1992.
196 pp. 1992 [3-540-55851-9]

Vol. 183: Hosoe, S. (Ed.)
Robust Control. Proceesings of a Workshop held in Tokyo, Japan, June 23-24, 1991.
225 pp. 1992 [3-540-55961-2]

Vol. 184: Duncan, T.E.; Pasik-Duncan, B. (Eds.)
Stochastic Theory and Adaptive Control. Proceedings of a Workshop held in Lawrence, Kansas, September 26-28, 1991.
500 pages. 1992 [3-540-55962-0]

Vol. 185: Curtain, R.F. (Ed.); Bensoussan, A.; Lions, J.L.(Honorary Eds.)
Analysis and Optimization of Systems: State and Frequency Domain Approaches for Infinite-Dimensional Systems. Proceedings of the 10th International Conference, Sophia-Antipolis, France, June 9-12, 1992.
648 pp. 1993 [3-540-56155-2]

Vol. 186: Sreenath, N.
Systems Representation of Global Climate Change Models. Foundation for a Systems Science Approach.
288 pp. 1993 [3-540-19824-5]

Vol. 187: Morecki, A.; Bianchi, G.; Jaworeck, K. (Eds.)
RoManSy 9: Proceedings of the Ninth CISM-IFToMM Symposium on Theory and Practice of Robots and Manipulators.
476 pp. 1993 [3-540-19834-2]

Vol. 188: Naidu, D. Subbaram
Aeroassisted Orbital Transfer: Guidance and Control Strategies.
192 pp. 1993 [3-540-19819-9]

Vol. 189: Ilchmann, A.
Non-Identifier-Based High-Gain Adaptive Control.
220 pp. 1993 [3-540-19845-8]

Vol. 190: Chatila, R.; Hirzinger, G. (Eds.)
Experimental Robotics II: The 2nd International Symposium, Toulouse, France, June 25-27 1991.
580 pp. 1993 [3-540-19851-2]

Vol. 191: Blondel, V.
Simultaneous Stabilization of Linear Systems.
212 pp. 1993 [3-540-19862-8]

Vol. 192: Smith, R.S.; Dahleh, M. (Eds.)
The Modeling of Uncertainty in Control Systems.
412 pp. 1993 [3-540-19870-9]

Vol. 193: Zinober, A.S.I. (Ed.)
Variable Structure and Lyapunov Control
428 pp. 1993 [3-540-19869-5]

Vol. 194: Cao, Xi-Ren
Realization Probabilities: The Dynamics of Queuing Systems
336 pp. 1993 [3-540-19872-5]

Vol. 195: Liu, D.; Michel, A.N.
Dynamical Systems with Saturation Nonlinearities: Analysis and Design
212 pp approx. 1994 [3-540-19888-1]